T0215292

SCIENCE IN THE MEDIA

This timely and accessible text shows how portrayals of science in popular media—including television, movies, and social media—influence public attitudes around messages from the scientific community, affect the kinds of research that receive support, and inform perceptions of who can become a scientist.

The book builds on theories of cultivation, priming, framing, and media models while drawing on years of content analyses, national surveys, and experiments. A wide variety of media genres—from Hollywood blockbusters and prime-time television shows to cable news channels and satirical comedy programs, science documentaries and children's cartoons to Facebook posts and YouTube videos—are explored with rigorous social science research and an engaging, accessible style. Case studies on climate change, vaccines, genetically modified foods, evolution, space exploration, and forensic DNA testing are presented alongside reflections on media stereotypes and disparities in terms of gender, race, and other social identities. *Science in the Media* illuminates how scientists and media producers can bridge gaps between the scientific community and the public, foster engagement with science, and promote an inclusive vision of science, while also highlighting how readers themselves can become more active and critical consumers of media messages about science.

Science in the Media serves as a supplemental text for courses in science communication and media studies, and will be of interest to anyone concerned with publicly engaged science.

Paul R. Brewer is a Professor of Communication and of Political Science & International Relations at the University of Delaware, USA.

Barbara L. Ley is an Associate Professor of Women & Gender Studies and of Communication at the University of Delaware, USA.

SCIENCE IN THE MEDIA

Popular Images and
Public Perceptions

Paul R. Brewer and Barbara L. Ley

Routledge
Taylor & Francis Group

NEW YORK AND LONDON

First published 2022
by Routledge
605 Third Avenue, New York, NY 10158

and by Routledge
2 Park Square, Milton Park, Abingdon, Oxon, OX14 4RN

Routledge is an imprint of the Taylor & Francis Group, an informa business

Library of Congress Cataloging-in-Publication Data
A catalog record for this title has been requested

ISBN: 978-1-032-04139-1 (hbk)
ISBN: 978-1-032-03399-0 (pbk)
ISBN: 978-1-003-19072-1 (ebk)

DOI: 10.4324/9781003190721

Typeset in Bembo
by KnowledgeWorks Global Ltd.

To Xander and Jingjing

CONTENTS

LIST OF FIGURES

ACKNOWLEDGEMENTS

We thank Brian Eschrich, Grant Schatzman, and everyone else at Routledge who worked on this book. We also thank our research collaborators—including James Bingaman, Natalie Jankowski, D. J. McCauley, Jessica McKnight, Emily Marquardt, Lucy Obozintsev, Ashley Painstil, Clint Townson, and David Wise—along with all the students who helped conduct the studies described in the book.

We're grateful to the University of Delaware and the University of Wisconsin-Milwaukee for supporting our research. In particular, we thank UD's College of Arts & Sciences, Department of Communication, Department of Women & Gender Studies, and Department of Political Science & International Relations, as well as UWM's Department of Journalism, Advertising, and Media Studies.

We thank the UD Center for Political Communication for funding many of the studies presented in the book. In addition, we're grateful to the UWM Institute for Survey and Policy Research, the Cultural Indicators Project, and Nancy Signorielli for providing us with data.

This book would not be possible without our research participants. We appreciate the time and effort they devoted to taking part in our studies.

Finally, we thank our parents for their support and our children for getting us to watch some of the movies and television shows that inspired this book.

ABOUT THE AUTHORS

Paul R. Brewer is a professor of Communication and of Political Science & International Relations at the University of Delaware. His research has appeared in journals such as *Public Opinion Quarterly*, *Public Understanding of Science*, and *Science Communication*, as well as magazines such as *National Geographic* and *Skeptical Inquirer*. He is also the author of *The Spirit Hollows*, a young adult novel about an inventor and an assistant undertaker who hunt spirits in a steampunk Appalachia.

Barbara L. Ley is an associate professor of Women & Gender Studies and of Communication at the University of Delaware. She is the author of *From Pink to Green: Disease Prevention and the Environmental Breast Cancer Movement* (Rutgers University Press, 2009). Her research has also appeared in journals such as the *Journal of Computer-Mediated Communication*, *Medical Anthropology*, *Public Understanding of Science*, *Science Communication*, and *Social Media + Society*. As a certified yoga instructor, she has developed trauma-informed programs for children, families, and undergraduate students.

1

IMAGES OF SCIENCE AND SCIENTISTS

Let's start with a picture.

Before you read any further, find a pen or pencil. Then draw a scientist (you can use the space below or a separate sheet of paper).

When you're finished, turn the page.

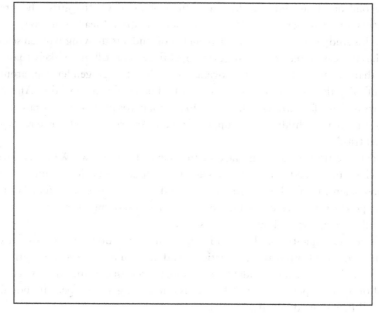

FIGURE 1.1 The reader's drawing of a scientist

DOI: 10.4324/9781003190721-1

Congratulations! You've just completed a test—known, logically enough, as the Draw-A-Scientist Test—that researchers have been using for more than half a century to study what people think scientists look like. The scholar who first developed it, David Wade Chambers, collected drawings of scientists from almost 5,000 elementary school students between 1966 and 1977.[1] Since then, researchers have administered the test to students at every education level, from kindergarten to college, as well as to teachers and teachers-in-training.[2]

Now let's take a closer look at your drawing. First question: how many of the following features does it include?

- Eyeglasses
- Wild or wacky hair
- A lab coat
- Pencils or pens in a pocket, or a pocket protector
- Beakers, test tubes, or other laboratory equipment

All of these are common features in young people's drawings of scientists.[3] The more of them you added to your picture, the closer it is to the "standard image"—or popular stereotype—of a scientist.[4] Children tend to learn this image over time, with older students including more stereotypical features in their drawings than younger ones do.[5]

Second question: did you draw a woman or man? Of the girls who drew scientists for Chambers, only 1% drew women scientists. Meanwhile, *every single boy* in his study drew a man. The percentage of students drawing women scientists has increased since the 1970s; even so, children are still more likely to draw men than women.[6] Beneath this overall trend lies a large gender gap: around half of all girls now draw women scientists, but far fewer boys do.[7] Another pattern is that older students are more likely than younger ones to draw men, suggesting that as children grow up, they tend to internalize gender stereotypes of scientists.[8]

Next question: what is the race of the scientist you drew? White children overwhelmingly tend to draw white scientists.[9] Students of color are more likely to draw scientists of color, but many of them draw white scientists, too.[10] Just as young people's pictures sometimes reflect gender stereotypes, these pictures can also reflect stereotypes about race and science.

Three more questions: How old is the scientist you drew? Does this person resemble anyone from real life or fiction? And did you draw a scientist with any visible disabilities? Many students draw older scientists. Some draw scientists based on specific people, with Albert Einstein serving as an especially popular model.[11] Few draw disabled scientists.[12]

The student drawing in Figure 1.2 includes every aspect of the popular stereotype of a scientist as captured by five decades of research. It depicts a man with wild hair and glasses, wearing a lab coat with a pocket protector and standing next to a table of beakers and flasks. He appears to be white. He is relatively old (judging by his hairline), bears something of a resemblance to Albert Einstein (with his mustache and hair), and has no visible disabilities. To many young people, *this* is what a scientist looks like.

So, who teaches students to stereotype—or, in some cases, not stereotype—scientists? For that matter, who shapes all the other perceptions that children *and* adults hold about science? Take a moment to consider whether you would agree with the following statements, and why:

- Scientists tend to be odd and peculiar people.
- Scientific work is dangerous.
- Scientists will probably bring an extinct species back to life by 2050.

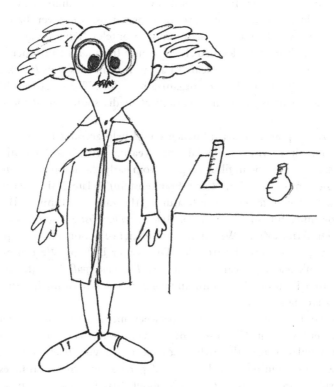

FIGURE 1.2 A stereotypical drawing of a scientist (an anonymous student's drawing)

- Most scientists agree that the earth is getting warmer because of human activity.
- Women who work in science are likely to experience bias based on their gender.
- Forensic DNA testing is a reliable form of evidence in criminal investigations.
- Paranormal investigators are scientific.

A host of messengers—including parents, schools, peers, politicians, religious leaders, and scientists themselves—could influence what we think about such topics. Given how much time most of us spend consuming media, however, our views of science may also mirror portrayals of it in Hollywood movies and prime time television; on cable news and satirical comedy programs; in science documentaries and children's cartoons; in Instagram selfies and YouTube videos.

The results of a national survey conducted by the Pew Research Center in 2017 illustrate the extent to which Americans depend on the media for whatever information they receive about science.[13] A majority of those polled (54%) said they learned about science from general news outlets, and almost half (45%) said the same of documentary science programs or videos. Moreover, half (49%) of the respondents watched science fiction movies or television programs. By comparison, only a third (33%) learned about science from family members, friends, or acquaintances, and far fewer learned about it from museums (11%), government agencies (10%), or advocacy organizations (6%). These patterns highlight the potential for messages from a variety of media sources to shape how the public perceives science.

As a case in point, media influence could help account for why so many young people draw white, male, lab coat-wearing scientists. After all, we can find this same stereotype in films ranging from early classics such as *Frankenstein* (featuring the misguided Dr. Henry Frankenstein) to late 20th century blockbusters such as *Back to the Future* (featuring the zany Dr. Emmett Brown) to recent box office hits such as *Captain America: The First Avenger* (featuring the weaselly Dr. Arnim Zola). We can find it on science fiction television programs such as *Fringe* (featuring the troubled Dr. Walter Bishop). We can even find it in children's television cartoons such as *The Powerpuff Girls* (featuring the good-natured Professor Utonium) and *Phineas and Ferb* (featuring the hapless Dr. Heinz Doofenshmirtz).

In this book, we explore the links between media portrayals of science and public perceptions of it. We look at many types of media, though we focus on the most popular ones of the early 21st century: movies, television, and social media. We also examine a wide range of perceptions, from estimates of the gender balance in astronomy to beliefs about "fringe" science topics such as extra-sensory perception (ESP) and ghost hunting. To understand *why* and *how* the media might shape public perceptions of science, we draw on four major theories of media effects. To capture *what* messages about science the media present, we combine case studies with "big picture" data. To test *whether* people's views reflect these portrayals, we use polls and experiments. Along the way, we consider how shifts in media messages could reshape public perceptions of science and scientists—as well as how audience members might go about assessing the messages they encounter.

Why Public Perceptions of Science and Scientists Matter

But first, we should explain why it's so important to understand people's perceptions of science and scientists. The short version is that these perceptions help set the stage for what scientific research takes place, who conducts it, who participates in it, and how it affects both public policy and individual behaviors.

To begin with, people's beliefs about science shape how willing they are to support scientific endeavors, including public funding for science.[14] Many university-based scientific projects depend on grants from governmental agencies such as the National Science Foundation (NSF) and the National Institute for Health (NIH). Numerous federal agencies, including the Environmental Protection Agency (EPA), the Centers for Disease Control and Prevention (CDC), and the National Oceanic and Atmospheric Administration (NOAA), conduct extensive research of their own, as well. In light of this, scientists and everyone else who values science should care about public support for it—particularly given efforts by some political leaders to slash federal funding for scientific research. To defend science, organizations such as the Union of Concerned Scientists and activist movements such as the March for Science have taken their cases directly to the public. So have federal agencies; for example, the National Aeronautics and Space Administration's (NASA) public relations efforts include everything from consulting on Hollywood films to posting photographs on Instagram.

Beyond funding, the image of science influences public opinion about a host of important policy issues. For example, trust in scientists—or the lack thereof—shapes people's beliefs about whether global warming is happening, whether the health benefits of vaccines outweigh their risks, whether humans evolved from other forms of life, whether genetically modified foods are safe to eat, and whether social distancing is necessary during pandemics.[15] Elected officials and other policymakers, in turn, sometimes (though not always) take public opinion into account when making decisions such as whether to take action on climate change, require that children receive vaccinations, mandate the teaching of evolution in schools, require labels for genetically modified foods, or mandate mask-wearing to prevent the spread of infectious diseases.[16]

On a more personal scale, perceptions of science matter for whether—and, if so, how—ordinary people rely on it in their everyday lives. Americans' views of science shape their own actions across many areas, including choices about whether to vaccinate their children, buy energy-efficient vehicles, and wear face masks during a pandemic.[17] Likewise, Americans' perceptions of science guide their participation on a range of issues that affect their own communities, from protecting local water supplies to choosing sites for nuclear reactors.[18]

The success of the growing drive for "citizen science" also depends on public perceptions of the scientific community. Recent years have witnessed a push for a model of science in which laypeople help produce scientific knowledge and engage in civic action alongside scientists.[19] Citizens who believe that scientists care about justice and listen to laypeople may welcome the chance to take part in scientific research and science-based decision-making.[20] By contrast, people who view scientists as distant or condescending may shy away from opportunities to participate in science.

Last, but not least, the public image of science can shape whether students pursue STEM (science, technology, engineering, and math) education and careers. Young people who view science as a force for good in society may decide that they want to grow up to scientists; meanwhile, those who perceive scientific work as dangerous and isolated—and scientists as strange and distant—may avoid a life of science.[21] Moreover, public perceptions of science could shape *who* becomes a scientist. For example, gender stereotypes about science may discourage young women from seeing themselves as potential scientists and exacerbate gender-based discrimination within scientific hiring and promotion.[22] Similarly, other sorts of stereotypes may create barriers for people of color, LGBTQ people, and disabled people who want to pursue scientific education and careers.[23]

What the Public Thinks about Science and Scientists

Against this background, many members of the scientific community worry that science has an "image problem"—one that could erode support for science funding, dampen trust in scientific conclusions, discourage public engagement with science, and hamper efforts to recruit scientists who reflect a diverse population. To be sure, a critical skepticism toward science is healthy for both science and society. As recent "replication crises" in fields such as nutrition and psychology illustrate, the scientific method requires researchers to question findings and retest hypotheses.[24] Furthermore, unethical and fraudulent practices on the part of individual researchers and broader scientific institutions highlight the need for oversight: consider the racist Tuskegee syphilis study (where medical investigators deceived Black participants) and Andrew Wakefield's retracted *Lancet* paper (where the author falsely claimed a link between autism and vaccinations based on fraudulent and unethical research).[25] At a deeper level, political, cultural, and economic forces shape what topics scientists choose to study, what methods they use, and how they interpret their findings, as well as what sorts of projects receive funding and whose voices are heard within the profession.[26]

In contrast to thoughtful skepticism, public alienation from or cynicism toward science could have destructive effects on science and society. A decade ago, science writers Chris Mooney and Sheril Kirshenbaum argued in their book *Unscientific America* that the "world of science ... can appear baffling, intimidating, and even downright unfriendly" to non-scientists.[27] Today, some observers see the public as increasingly hostile to science. For example, a 2017 *Scientific American* article by Tom Nichols opens with the question, "Do Americans hate science?" and then answers, "They certainly seem to hate it more than they used to, as they rage against experts in every field."[28]

So, how *do* people perceive science and scientists? As it turns out, quite positively—at least in general terms. Surveys conducted by the NSF and National Opinion Research Center (NORC) from 1979 to 2018 show that around 70% of Americans have consistently agreed that the benefits of scientific research outweigh any harmful results from it.[29] Public confidence in science has also remained steady since the 1970s even as faith in most institutions has declined.[30] Of the people polled by NORC in 2018, 44% said they had a great deal of confidence in the scientific community, and another 47% had some confidence in it. Organized religion, big business, the press, and the three branches of government all scored much lower.

When it comes to public funding for science, people's perceptions are similarly positive. From the 1980s to the present, around three-fourths of the respondents in NSF and NORC polls have consistently agreed that the government should fund basic scientific research. Over the same time span, the proportion saying that spending on scientific research is "about right" or "too little" has grown from a large majority (78%) to a slightly larger one (83%). Americans *do* rank other spending priorities—including education, health, and law enforcement—higher, but few want to cut funding for scientific research.

General perceptions of scientists are overwhelmingly—and increasingly—positive, too. Of the respondents in a 2001 NSF survey, 86% agreed that scientists work for the good of humanity, with 11% strongly agreeing. Fifteen years later, an NORC survey found that 89% of the public agreed with the same statement—and that strong agreement had jumped to 26%. We came away with a nearly identical picture when we conducted three nationally representative polls of our own in July 2016, October 2016, and October 2018. Around 90% of the respondents in each survey perceived scientists as working for good.[31]

On the other hand, a large and growing percentage of Americans see scientists as strange, if not downright arrogant. Almost a quarter of the respondents in the 2001 NSF survey (24%) agreed that scientists tend to be odd and peculiar. By 2016, that figure had doubled to around half the respondents in the NORC survey (52%). Our own surveys tell essentially the same story. The percentage of respondents saying that scientists tend to be odd and peculiar was 42% in our July 2016 poll, 47% in our October 2016 poll, and 39% in our October 2018 poll. Furthermore, almost a third of the respondents in the October 2016 survey (32%) said that scientists tend to look down on other people.

Many people also hold distorted perceptions about the nature of scientific work. Around half the people we polled—47% of those in the July 2016 survey, 50% in the October 2016 survey, and 45% in the October 2018 survey—saw scientific work as dangerous. In addition, substantial percentages of respondents—30% in July 2016, 30% in October 2016, and 22% in October 2018—thought that scientists usually work alone. Such results may come as a surprise to the vast majority of scientists who collaborate with one another on innocuous tasks rather than secluding themselves in remote castles or on secret islands to build killer robots or genetically engineer dangerous new life forms that wreak havoc.

All of this suggests that the public's view of science is not so much negative as it is largely-positive-but-somewhat-leery. Most people see science as beneficial to society and scientists as working to help others, yet many see scientists as weird and scientific work as a scary, solitary endeavor. The first set of attitudes could foster support for scientific endeavors, but the second set may discourage some Americans from participating in science themselves.

As for the origins of these perceptions, media messages could shape both the positive and not-so-positive ones. Movies, television shows, and other outlets often portray scientists as sympathetic and even heroic. Think of Mark Watney, the intrepid botanist played by Matt Damon in the 2015 film *The Martian*, or the crime-solving forensic scientists on the prime-time television dramas *CSI: Crime Scene Investigation* and *Bones*. At the same time, popular media also offer stereotypical images of oddball scientists such as Sheldon Cooper from the television sitcom *The Big Bang Theory*, along with stories of scientists being hurt or killed by supervillains, aliens, and monsters on science fiction shows such as *Lost* and *Marvel's Agents of S.H.I.E.L.D.*

Bridging Divides and Fostering Engagement

Another concern within the scientific community is that members of the US public are now, more than ever, rejecting what scientists conclude about important questions. In an April 19, 2017 segment of his program *StarTalk*, astrophysicist Neil deGrasse Tyson laments that, "People have lost the ability to judge what is true and what is not, what is reliable, what is not reliable … That's not the country I remember growing up in. I don't remember any other time where people were standing in denial of what science was." Along the same lines, popular science educator Bill Nye "the Science Guy" warns in a 2018 *POV* documentary that "if we raise a generation of kids [who] … can't think scientifically, we are headed for trouble … These people who are denying science, denying evolution, denying the efficacy of vaccinations, and especially denying human-caused climate change—we just can't have this."

Though Americans trust science in general terms, scientists and the public *are* far apart on some of the most crucial scientific issues facing the country. When the Pew Research Center surveyed both the US public and members of the American Association for the Advancement of Science (AAAS) in 2014, the results showed wide gaps on a range of topics.[32] Among scientists, there was an overwhelming consensus that humans have evolved over time (98% of the AAAS members surveyed agreed with this), that it's safe to eat genetically modified foods (88%), that climate change is mostly due to human activity (87%), and that childhood immunizations such as the MMR (measles-mumps-rubella) vaccine should be required (86%). In contrast, only two-thirds of the public thought that vaccines should be required (68%, for a gap of 18 points between scientists and the public) and that humans have evolved (65%, for a 33-point gap). Only half the public believed that climate change is mostly human-caused (50%, for a 37-point gap), and barely more than a third saw genetically modified foods as safe (37%, for a 51-point gap).

This is not to say that scientists and the public differ on everything. In the Pew surveys, for example, most scientists *and* most members of the public agreed that the International Space Station has been a good investment for the United States (68% and 64%, respectively) and that we should increase our use of bioengineered fuels (78% and 68%). Meanwhile, neither group (only 39% of scientists and 31% of the public) tended to favor the increased use of fracking to extract oil and natural gas from underground rock formations. Still, the Pew results highlight the stark divides between scientists and the public on a host of key subjects.

One potential explanation for such gaps is that scientific illiteracy prevents members of the public from grasping scientists' conclusions. This premise underlies what researchers call the *deficit model* of science communication.[33] At first glance, the model seems to fit with results from pop quizzes showing how much—or, in many cases, how little—ordinary people know about science. To test your knowledge against the public's, try answering the following questions from a 2016 Pew Research Center survey:[34]

- Which gas makes up most of Earth's atmosphere—carbon dioxide, hydrogen, nitrogen, or oxygen?
- Which of these terms refers to health benefits occurring when most people in a population get a vaccine—herd immunity, population control, or vaccination rate?
- Humans and mice share the same genetic makeup by: less than 10%; between 11% and 49%, or about 50% or more?

The correct answers are "nitrogen," "herd immunity," and "about 50% or more." If you missed any of them, don't feel bad: for each question, only around a third of the public gave the right answer. Looking at such knowledge deficits, it's tempting to argue that so many people disagree with the scientific consensus on topics such as climate change, vaccines, evolution, and genetically modified food because they simply don't understand the science involved.

Just as the deficit model offers a seemingly straightforward explanation for the divides between scientists and the public, it also suggests a way to bridge them: namely, working harder to deliver information from the scientific community to laypeople. If the problem is scientific illiteracy, then the obvious solutions are better science education and more efforts by scientists to enlighten the public. In other words, the best way to get the public to agree with scientists is for the latter to "science at" the former more often and more loudly. Accordingly, outreach efforts based on the deficit model typically follow a top-down, one-way communication format, where scientists try to explain their findings to an audience assumed—either explicitly or implicitly—to consist of passive, empty vessels waiting to be filled with facts and findings.[35]

There is a logic to this approach: scientific knowledge *can* influence how members of the public form their perceptions of science-related topics.[36] Yet its impact is often limited. For example, the Pew Research Center's 2016 survey showed that respondents' scores on its science quiz were "only modestly and inconsistently" linked to their belief in human-caused climate change. In many cases, what scientists themselves say about science may matter less than what other messengers, including the media, say about it.[37]

To complicate matters, different media outlets may sway their audiences in different ways depending on how they cover scientific issues. Consider climate change, which has emerged as not only a high-profile scientific concern but also a major partisan fault line in US politics. For this issue, the impact of watching a conservative-leaning cable television news network such as the Fox News Channel could vary dramatically from the impact of watching the liberal-leaning MSNBC cable network, a satirical television news program such as Comedy Central's *The Daily Show*, or a Hollywood film such as *The Day After Tomorrow*.

The deficit model breaks down even further when we consider the active roles audience members play in choosing media sources and interpreting media messages. Indeed, the *public understanding of science model* highlights how people make sense of media messages based on their own experiences, values, interests, and life stages.[38] As such, it emphasizes two-way dialogue, rather than top-down transmission of information, as a foundation for constructive communication between scientists and laypeople. Following this model, we argue that scientists and other science communicators should craft messages intentionally and strategically based on who their audience members are and what worldviews they hold. We also build on the model's premise that citizen participation in scientific research and decision-making benefits both society and science itself.[39] In particular, we explore how media messages can foster broad and critical-minded engagement with science.

Biases and Underrepresentation in Science

Along with the challenges of bridging divides with laypeople and fostering engagement, the scientific community faces another important task: attracting and retaining a more diverse workforce. In recent years, scientific associations such as the AAAS, leading journals such as *Scientific American* and *Nature*, and federal agencies such as NASA and the NIH have all called for strengthening efforts to recruit scientists from underrepresented groups.[40] The leaders of these organizations point to three reasons for doing so. First, promoting equal opportunity in science is the fair, and right, thing to do. Second, ensuring that people from all backgrounds are welcome and valued in science will keep us from missing out on a huge pool of talented potential scientists. Third, bringing people from different backgrounds together to collaborate fosters greater creativity, more innovation, and better problem solving—critical elements for successful scientific research.

The scientific community's history in terms of diversity leaves much room for improvement. Women, people of color, LGBTQ people, and disabled people have long been underrepresented in STEM professions. To cite one example: in the first 117 years of the Nobel Prize for physics, a grand total of 2 women received the award—Marie Curie in 1903 and Maria Goeppert-Mayer in 1963.[41] And another example: only 40 Black students earned doctorates in astronomy or astrophysics between 1955 and 2012, for an average of fewer than 1 a year (that figure includes Neil deGrasse Tyson, who went on to become the director of the Hayden Planetarium and a leader in the drive to demote Pluto from planet status).[42]

Despite advances in some areas, these patterns have persisted through the 2010s. Consider the National Science Board's numbers from 2017.

- Women, who made up slightly more than half the population, accounted for only 29% of the US science and engineering workforce (a mere 6-point increase from 1993).
- Only 6% of the people working in science and engineering were Black and only 8% were Hispanic (by comparison, Black and Hispanic people accounted for 13% and 16% of the 2010 US population, respectively).[43]:

Similarly, a 2013 survey of federal workers found that LGBTQ people accounted for only 80% of the STEM jobs one would've expected going by population figures, and 2015 data show that people with disabilities occupied only two-thirds of the science and engineering jobs one would've expected based on their numbers in the overall US population.[44]

Furthermore, many of the women, people of color, LGBTQ people, and disabled people who do pursue STEM careers encounter hostility, invisibility, stereotyping, and implicit bias along with the absence of supportive institutions, policies, accommodations, and mentors. A 2017 Pew Research Center survey found that half of all women working in STEM jobs had experienced gender discrimination at work, and more than a third (36%) said that sexual harassment was a problem in their workplace.[45] The same survey found that 62% of Black STEM workers, 44% of Asian American STEM workers, and 42% of Hispanic STEM workers had experienced discrimination based on their race or ethnicity.[46] Likewise, other studies have shown that LGBTQ people and disabled people face multiple barriers in scientific professions.[47]

Such obstacles can also reinforce one another. In a landmark 1976 report, AAAS researchers Shirley Mahaley Malcom, Paula Quick Hall, and Janet Welsh Brown observed that women of color in STEM faced a "double bind" of gender-based discrimination *and* race or ethnicity-based discrimination."[48] Though almost half a century has passed since then, these mutually reinforcing barriers remain in place.[49] Similar double-binds confront people of color who are LGBTQ or who have disabilities.[50] Even triple-binds are possible: for example, women of color who are LGBTQ or who have disabilities face multiple intersecting barriers in STEM.[51]

Yet not everyone perceives the presence of these obstacles. In the case of gender bias, our July 2016 survey yielded a divided picture on whether the wider public sees science as a welcoming or hostile environment for women. On the one hand, a sizable majority of those we polled (62%) agreed that women who work in science are likely to experience bias based on their gender. On the other hand, around the same proportion (65%) said that women have the same opportunities as men to succeed in science (which means that some respondents thought *both* statements were true). We found a sizable gender gap on each question, as well. Compared to men, women were 11 percentage points more likely to perceive gender bias in science and 14 percentage points less likely to perceive equal opportunities.

All of which raises the question: do scientists and laypeople support efforts to make science more diverse? We found high levels of support when we asked respondents in our October 2018 survey whether they agreed that "we should do more to promote" four different aspects of diversity in science. Of the respondents, 89% favored promoting "opportunities for women to work in science," 89% favored promoting "opportunities for people with disabilities to work in science," 82% favored promoting "racial and ethnic diversity among people working in science," and 76% favored promoting "opportunities for gay men, lesbians, and transgender people to work in science." However, it's important to remember that these questions measured support in the abstract, rather than in terms of specific measures or cases.

It's also worth mentioning that support for promoting diversity in science differed depending on who was being asked. For example, women were more likely than men to agree with promoting opportunities for women in science (93% versus 83%). Likewise, people of color were more likely than white respondents to agree with promoting racial and ethnic diversity in science (87% versus 81%), just as LGBTQ respondents were more likely than non-LGBTQ respondents to agree with promoting opportunities for gay men, lesbians, and transgender people in science (81% versus 75%).

Although success in diversifying science depends in part on views among both scientists and the broader public, it depends on the perceptions held by potential future scientists, too. Young people's decisions about whether to pursue educational opportunities and careers in science reflect their own views of the profession and the people working in it; as a result, group differences in these perceptions may contribute to the long-standing demographic disparities in science. Take the gender gap in the scientific workforce. Compared to young men, young women tend to report lower science interest, greater anxiety about science, and greater perceptions of gender bias in science. Such "belief gaps" may serve as barriers to entry into the profession.[52]

By shaping audience members' perceptions of science and scientists, media portrayals could either reinforce existing biases and underrepresentation or help promote greater diversity. For example, The Big Bang Theory's initial focus on four men working as physicists or engineers and one woman working as a restaurant server/aspiring actor may have sent the implicit message that STEM is a "boy's club," while the same program's subsequent introduction of two women biologists may have tempered that message. Depictions of scientists on children's programs—including not only educational ones such as The Magic Schoolbus and Mission Unstoppable but also entertainment-oriented ones such as Phineas and Ferb—could likewise echo or challenge stereotypes of scientists as white men. Meanwhile, YouTube channels such as The Brain Scoop with Emily Graslie may influence social media users' perceptions by directly calling out biases in science.

Explaining Media Effects

Four leading theories of media effects point to specific routes by which the media could influence public views of science and scientists. The first of these is cultivation theory, originally developed by communication scholar George Gerbner. He argued that people's perceptions of society come to reflect the pictures of it that dominate the media—and that the more media messages people consume, the more likely they are to think the real world resembles the media world.[53] By his theory's logic, a steady stream of media images depicting botanists or geologists performing heroic deeds might foster positive perceptions of scientists whereas frequent portrayals of astronomers being killed by aliens or paleontologists being eaten by dinosaurs might sway people to see science as a hazardous occupation. Gerbner himself emphasized the role of television in cultivating perceptions of reality, given that this medium became the leading source of cultural images starting in the mid-20th century and has remained so ever since.[54] Yet the logic of cultivation theory could apply to any medium, from movies to social media.

Whereas cultivation theory addresses the effects of audience members' long-term exposure to the dominant messages within the entire media system, priming theory—as adapted from psychology to media effects research by scholars such as Shanto Iyengar and Donald Kinder—explains how people mentally process individual pieces of new information. The central premise behind priming is that we, as humans, can only think so hard about so many things at once.[55] As a result, we tend to base our impressions on the thoughts that come to mind most readily. In terms of media effects, this means that people tend to be influenced by whatever messages they've received most often and most recently. For example, watching a television show about shark attacks on The Discovery Channel could prime audience members to dwell on their fear of such attacks when forming opinions about shark conservation efforts.[56] Likewise, mentioning the television crime drama *CSI* to viewers could prime them to think about the show's portrayals of forensic science when considering the reliability of DNA evidence.[57]

Just as priming involves media messages shaping *what* audience members think about, framing involves the impact of such messages on *how* audience members think about a topic. Framing theory, which builds on the work of sociologists such as Erving Goffman and William Gamson along with psychologists such as Daniel Kahneman and Amos Tversky, suggests that the ways in which the media tell a story can influence audience members' interpretations of it.[58] In framing a person, event, or issue, any messenger—be they filmmaker, journalist, comedian, or ordinary citizen—decides what the story is "all about": which aspects to emphasize and which to omit; what language, images, symbols, and metaphors to use; whom to assign credit or blame.[59] As a case in point, a news story about genetically modified foods could frame them as examples of scientific progress or as dangerous "Frankenfoods."[60] Likewise, a television comedy sketch about climate change could frame it in terms of an evenly balanced debate or an overwhelming scientific consensus on one side, and a reality show about paranormal investigators could frame them as scientists or pseudoscientists.[61] These storytelling choices, in turn, may sway whether audience members perceive genetically modified foods as safe to eat, or believe in human-caused global warming, or view ghost hunters as credible.

Our final media effects theory, social cognitive theory, helps explain people's decisions about who they want to be like. First proposed by psychologist Albert Bandura and applied to science communication by scholars such as Jocelyn Steinke and Marilee Long, this theory posits that people learn by observing and identifying with cultural models, including ones they see in the media.[62] In particular, children (who spend several hours a day on average watching screens) can form a range of attitudes, including their career aspirations and their beliefs about gender roles, through "vicarious contact" with the people they see in movies, television shows, or online videos.[63] Accordingly, young viewers who identify with real-life scientists and science educators in the media—such as *Cosmos* host Neil deGrasse Tyson, *Brain Scoop* host Emily Graslie, or the eponymous host of *Bill Nye the Science Guy*—may wind up picturing themselves as potential scientists. The same could be true for children and adolescents who see their own "possible selves" in fictional scientists such as Jane Foster from the *Thor* movies or Princess Bubblegum from the animated children's show *Adventure Time*.[64]

In focusing on the effects of media messages about science, we don't mean to suggest that audience members come to such messages as blank slates. Far from it: people draw on a variety of experiences and values in forming their perceptions of science and scientists. Take higher education, which can foster more positive attitudes toward science in general.[65] Or political conservatism and religiosity, both of which are associated with stronger reservations about science.[66] Members of the public often weigh these factors as heavily, or even more heavily than, the messages they receive from the media.

Nor do audience members passively accept everything they encounter in television, movies, and social media. Instead, they can use their prior beliefs in choosing whether to accept or reject media messages. For example, Republican viewers might tune out evidence for global warming presented by the progressive-leaning cable news network MSNBC, whereas Democratic viewers might do the same for dismissive coverage of climate change on the conservative-leaning Fox News Channel.[67] By the same logic, liberal audience members who watch late-night talk show host Stephen Colbert scoff at global warming could take his comments as ironic jokes, whereas conservative viewers could interpret the same statements as "funny because they're true."[68] With this in mind, we sometimes explore not only *whether*—and if so, *how*—the media influence public perceptions of science but also *which* media sources do so for *whom*.

Studying Media Messages and Media Effects

Our research looks at both the nature of media messages about science and the effects of those messages on audience members. In studying the messages, we draw on a series of case studies. For instance, our discussion of movies highlights two recent Hollywood blockbusters: *Jurassic World* and *The Martian*. Similarly, our analysis of prime time television uses *The X-Files* and *The Big Bang Theory* as centerpieces, while our examination of science documentaries revolves around *Shark Week*, *NOVA*, *MythBusters*, and the two versions of *Cosmos*. When we discuss media messages about fringe science, we focus on *Ghost Hunters*, *Ancient*

Aliens, and *Finding Bigfoot*. When we talk about children's media, we do the same for *Bill Nye the Science Guy*, *The Magic School Bus*, and *Phineas and Ferb*.

Along with these cases, we describe broader studies of media content, including ones exploring how prime time television depicts scientists; how science-related Facebook pages frame their posts; and how crime dramas portray the process of forensic DNA analysis. Our evidence here comes from research we and other scholars have conducted through content analysis, a method that involves sorting large quantities of media messages—such as character portrayals on fictional television shows, segments on cable television news, or jokes on late-night comedy shows—into different categories.[69] Researchers who use this technique typically assign multiple people to code a given set of messages and then check for a high level of agreement across the coders to ensure that the categorization process is reliable.

When it comes to studying the impact of media messages, we rely on two main tools: public opinion polls and randomized experiments. The key strength of the first method lies in its power to capture representative portraits of what the public thinks. So long as a poll's respondents are selected by scientific means (such as probability sampling), and so long as the poll's sample size is large enough, its results should accurately mirror what we would find if we surveyed the entire population.[70] In this book, we draw on data from many different polls of the US public—some conducted by us (as we describe in the Appendix), and some by other organizations; some conducted by telephone, and others conducted through the internet using nationally representative pools of respondents.

Although polling is our primary method for measuring public perceptions about science and scientists, there are challenges in using it to test what shapes such perceptions. One approach to studying media effects would involve simply asking people whether—and, if so, how—media messages influence their views. For example, a 2017 Pew Research Center survey asked respondents whether watching three different types of shows and movies helped or hurt their understanding of science, technology, and medicine.[71] Among viewers of shows and movies about criminal investigations, 30% said "helps" and 11% said "hurts." For shows focused on hospital and medical settings, the percentages were 23% and 12%, respectively. Meanwhile, 13% of science fiction viewers said "helps" and the same percentage said "hurts."

Yet there is good reason to interpret such responses cautiously. For starters, some members of the public may be unwilling to admit to pollsters (or even themselves) that media messages shape their beliefs.[72] Furthermore, many survey respondents may not be entirely conscious of how they form their perceptions of science.[73] If people absorb ideas from television, movies, and social media without remembering that they did so, then their own assessments of what shapes their perceptions won't capture the full story.

Given the limitations of self-reports about media effects, researchers who use surveys to study such effects often rely on a second approach: correlational analysis, which involves testing for relationships between what media respondents consume and what they believe. Suppose we want to find out whether cable news coverage of climate change influences viewers' perceptions. Here, we could ask survey respondents one set of questions about their news habits and another set of questions about climate change, and then look for patterns between the first set of answers and the second. Specifically, we could analyze whether CNN or MSNBC viewing goes hand in hand with belief in human-caused global warming, as well as whether Fox News watchers are particularly inclined to express skepticism about climate change.

This approach has its own limits, however, when it comes to testing for media effects. Evidence that using a certain form of media is *correlated* with holding a certain belief isn't necessarily proof that the former *caused* the latter. In some cases, a third factor—or set of factors—could shape both media use and perceptions. As an illustration, suppose that watching Fox News and disbelieving in human-caused climate change only appear to go hand in hand because conservatives are more likely than liberals to do both. If so, then our evidence for a "Fox News effect" on global warming skepticism would be illusory.

One partial safeguard against drawing mistaken inferences from survey data involves statistically controlling for the impact of factors besides media use, such as political ideology and demographics, that might influence people's perceptions of science and scientists. But that still leaves another potential problem: if, after accounting for other "likely suspects," we find a relationship between using a specific form of media and holding a specific belief, how do we know which influenced which? In some cases, it may be more plausible that media use would affect perceptions of science than vice versa. For example, there probably aren't too many people who base their overall levels of television viewing on their impressions of scientists; as a result, any relationship between the two presumably reflects media influence rather than selective viewing. In other cases, interpreting relationships between media use and audience members' perceptions may be thornier. Suppose we find that people who watch paranormal-themed reality television programs are particularly likely to believe that houses can be haunted. Would this result suggest that the shows influence the beliefs or that the beliefs lead people to watch the shows (or maybe even a combination of the two)?

In such cases, we may need stronger evidence of cause and effect—evidence that our second approach to studying media influence can help supply. To capture the impact of specific messages, we and other media researchers sometimes conduct experiments that involve randomly dividing a pool of participants into groups and giving each group a different message (or no message). By comparing how the people in each group (or condition) answer questions on a posttest survey, we can then determine the impact of our treatments—that is, the messages.[74]

In looking at how media messages shape audience members' perceptions of science, we draw on results from a variety of randomized experiments, including ones testing the effects of clips from *The Big Bang Theory* on stereotypes of scientists, the effects of YouTube videos on perceptions of gender bias in science, and the effects of news stories on beliefs about ESP.

The major challenge in using this approach to study media effects involves making the leap from what happens in the experiment to what happens beyond it. If our participants are college students, then we should be careful when using our results to draw conclusions about the broader public. If we only test the impact of a few messages, then we should bear in mind that other messages may produce different effects. Most of all, we should remember that how people behave in the artificial setting of an experiment may not always be the same as how they behave in the real world during their day-to-day lives.

The overarching point here is that no single method is an ideal one for studying media messages or media effects, which is why we use a combination of research tools. Case studies give us detailed portraits of how specific media outlets depict science, while content analyses offer us windows onto the broader landscape of media messages. Polls capture representative snapshots of public opinion, while experiments provide stronger tests of media effects. By assembling evidence from all these approaches, we can come away with a deeper understanding of media messages, public perceptions, and the connections between the two.

Reshaping Media Messages and Public Perceptions

We've already asked you to draw your own picture of a scientist. We've also shown you a drawing that reflects the stereotype of a scientist as a white, wild-haired man wearing glasses and a lab coat, surrounded by test tubes and beakers. Now, let's look at a third picture. The student drawing in Figure 1.3 includes some stereotypical aspects, such as a lab coat and beakers, but departs from the "standard image" in other ways: *this* scientist is a woman who doesn't have glasses or wild hair. Though many children (and adults) still picture scientists as looking like Albert Einstein, or Dr. Frankenstein, or Doc Brown, present-day students—particularly girls and young women—are increasingly likely to draw scientists who reflect an alternative vision. When it comes to how we visualize scientists, the picture can change—literally.

Media images can evolve, too. Some television shows and movies still depict scientists in ways that reinforce popular stereotypes, but not every media portrayal fits the historical pattern. Recent films such as *Hidden Figures* and *Black Panther* offer a less stereotypical, more diverse portrait of the profession, as do television programs such as *Cosmos: A SpaceTime Odyssey*. And if media messages

FIGURE 1.3 Another drawing of a scientist (an anonymous student's drawing)

can influence how people view science, then changes in such messages could help reshape their mental images of scientists.

The same principle applies to other sorts of messages and perceptions about science. By looking at how the media *do* portray science and how those portrayals *do* influence audience members, we can also begin to learn how alternative portrayals *could* lead to different effects. Ultimately, we hope to illuminate ways in which media messengers can engage more constructively with the public. For scientists and science educators, a better understanding of media effects could aid in bridging divides with laypeople, fostering citizen science, and promoting diversity in the profession—all of which are important goals for the scientific community. For media producers, the same understanding could inform decisions about how to frame news stories, jokes, and social media posts about science, as well as how to portray scientists in ways that entertain audiences while avoiding harmful stereotypes.

Looking at how and when media messages influence audience members' views of science can also shed light on which strategies are effective for encouraging people to become more active, thoughtful consumers of these messages. Do televised public service announcements and social media fact-checks blunt the impact of scientific misinformation or reinforce it? Does media literacy training prepare students to resist stereotypical portrayals of science? Does including skeptical voices—or, alternatively, using humor—counteract spurious claims to scientific authority? Answering such questions is a crucial task if we want to foster critical-minded public engagement with science.

One last thought: we believe studying media messages about science is important for big-picture reasons, but we also think it's interesting for its own sake—not to mention fun. We've spent the past decade and a half researching the topic in no small part because we enjoyed watching *Ghostbusters*, *Star Trek*, and the original *Cosmos* miniseries as children. Because we enjoyed marathoning *CSI* during a trip to Las Vegas back in 2005 (an experience that inspired our first co-authored article). Because we've enjoyed watching *The Octonauts* and *Wild Kratts* with our own children, and because we've enjoyed catching up with *Marvel's Agents of S.H.I.E.L.D.* and *Stranger Things* after their bedtimes. This book is about media theories and findings, but it's also about dinosaurs and sharks; about detectives and time travelers; about Bigfoot hunters and Martians. For the parts about the dinosaurs and the Martians, we turn to our first medium of interest: Hollywood movies.

Notes

1 Chambers, David Wade, "Stereotypic images of the scientist: The Draw-A-Scientist Test," *Science Education* 67, no. 2 (1983): 255–265.
2 Finson, Kevin D., "Drawing a scientist: What we do and do not know after fifty years of drawings," *School Science and Mathematics* 102, no. 7 (2002): 335–345; Miller, David I., Kyle M. Nolla, Alice H. Eagly, and David H. Uttal, "The development of children's gender-science stereotypes: A meta-analysis of 5 decades of US Draw-a-Scientist studies," *Child Development* 89, no. 6 (2018): 1943–1955.
3 Chambers, "Stereotypic images"; Finson, "Drawing a scientist"; Miele, Eleanor, "Using the Draw-A-Scientist Test for inquiry and evaluation," *Journal of College Science Teaching* 43, no. 4 (2014): 36–40; Steinke, Jocelyn, Maria Knight Lapinski, Nikki Crocker, Aletta Zietsman-Thomas, Yaschica Williams, Stephanie Higdon Evergreen, and Sarvani Kuchibhotla, "Assessing media influences on middle school—aged children's perceptions of women in science using the Draw-A-Scientist Test (DAST)," *Science Communication* 29, no. 1 (2007): 35–64.
4 Chambers, "Stereotypic images." 256.
5 Chambers, "Stereotypic images."
6 Miller et al., "Gender-science stereotypes."
7 Miller et al., "Gender-science stereotypes."
8 Miller et al., "Gender-science stereotypes."
9 Finson, "Drawing a scientist."
10 Finson, "Drawing a scientist."

11 Miele, "Using the draw-a-scientist test."

12 None of the articles we examined included any drawings that depicted a scientist with a visible disability, nor any discussion of such drawings.

13 Funk, Cary, Jeffrey Gottfied, and Amy Mitchell, "Science news and information today," Pew Research Center, Sept. 20, 2017, www.journalism.org/2017/09/20/science-news-and-information-today.

14 Besley, John C., "The National Science Foundation's science and technology survey and support for science funding, 2006–2014," *Public Understanding of Science* 27, no. 1 (2018): 94–109; National Science Board, "The state of U.S. science and engineering 2020," National Science Foundation/National Science Board, 2020. https://ncses.nsf.gov/pubs/nsb20201.

15 Barry, Colleen, Hahrie Han, and Beth McGinty, "Trust in science and COVID-19," Johns Hopkins Bloomberg School of Public Health, June 17, 2020, www.jhsph.edu/covid-19/articles/trust-in-science-and-covid-19.html; Hamilton, Lawrence C., Joel Hartter, and Kei Saito, "Trust in scientists on climate change and vaccines," *Sage Open* 5, no. 3 (2015): 2158244015602752; Hmielowski, Jay D., Lauren Feldman, Teresa A. Myers, Anthony Leiserowitz, and Edward Maibach, "An attack on science? Media use, trust in scientists, and perceptions of global warming," *Public Understanding of Science* 23, no. 7 (2014): 866–883; Lang, John T., and William K. Hallman, "Who does the public trust? The case of genetically modified food in the United States," *Risk Analysis* 25, no. 5 (2005): 1241–1252; Nadelson, Louis S., and Kimberly K. Hardy, "Trust in science and scientists and the acceptance of evolution," *Evolution: Education and Outreach* 8, no. 1 (2015): 9.

16 Berkman, Michael B., and Eric Plutzer, "Scientific expertise and the culture war: Public opinion and the teaching of evolution in the American states," *Perspectives on Politics* 7, no. 3 (2009): 485–499; Bord, Richard J., Robert E. O'Connor, and Ann Fisher, "In what sense does the public need to understand global climate change?" *Public Understanding of Science* 9, no. 3 (2000): 205–218; Motta, Matthew, Timothy Callaghan, and Steven Sylvester, "Knowing less but presuming more: Dunning-Kruger effects and the endorsement of anti-vaccine policy attitudes," *Social Science & Medicine* 211 (2018): 274–281; Weiss, Barry D., and Michael K. Paasche-Orlow, "Disparities in adherence to COVID-19 public health recommendations," *HLRP: Health Literacy Research and Practice* 4, no. 3 (2020): e171–e173; Wohlers, Anton E., "Labeling of genetically modified food: Closer to reality in the United States?" *Politics and the Life Sciences* 32, no. 1 (2013): 73–84.

17 Gilkey, Melissa B., William A. Calo, Macary W. Marciniak, and Noel T. Brewer, "Parents who refuse or delay HPV vaccine: Differences in vaccination behavior, beliefs, and clinical communication preferences," *Human Vaccines & Immunotherapeutics* 13, no. 3 (2017): 680–686; Padilla, Maria, "Who's wearing a mask? Women, Democrats, and city dwellers," *New York Times*, June 2, 2020, www.nytimes.com/2020/06/02/health/coronavirus-face-masks-surveys.html; O'Connor, Robert E., Richard J. Bord, and Ann Fisher, "Risk perceptions, general environmental beliefs, and willingness to address climate change," *Risk Analysis* 19, no. 3 (1999): 461–471.

18 Besley, John C., "Public engagement and the impact of fairness perceptions on decision favorability and acceptance," *Science Communication* 32, no. 2 (2010): 256–280; Haywood, Benjamin K., and John C. Besley, "Education, outreach, and inclusive engagement: Towards integrated indicators of successful program outcomes in participatory science," *Public Understanding of Science* 23, no. 1 (2014): 92–106; VanDyke, Matthew S., and Andy J. King, "Using the CAUSE model to understand public communication about water risks: Perspectives from Texas groundwater district officials on drought and availability," *Risk Analysis* 38, no. 7 (2018): 1378–1389.

19 Bonney, Rick, Tina B. Phillips, Heidi L. Ballard, and Jody W. Enck, "Can citizen science enhance public understanding of science?" *Public Understanding of Science* 25, no. 1 (2016):

2–16; Brossard, Dominique, Bruce Lewenstein, and Rick Bonney, "Scientific knowledge and attitude change: The impact of a citizen science project," *International Journal of Science Education* 27, no. 9 (2005): 1099–1121.

20 Dudo, Anthony, and John C. Besley, "Scientists' prioritization of communication objectives for public engagement," *PloS One* 11, no. 2 (2016): e0148867; McComas, Katherine A., John C. Besley, and Zheng Yang, "Risky business: Perceived behavior of local scientists and community support for their research," *Risk Analysis* 28, no. 6 (2008): 1539–1552.

21 Besley, John C., "Predictors of perceptions of scientists: Comparing 2001 and 2012," *Bulletin of Science, Technology & Society* 35, no. 1–2 (2015): 3–15; Losh, Susan Carol, "Stereotypes about scientists over time among US adults: 1983 and 2001," *Public Understanding of Science* 19, no. 3 (2010): 372–382; National Science Board, "State of U.S. science."

22 Funk, Cary, and Kim Parker, "Women and men in STEM often at odds over workplace equity," Pew Research Center, Jan. 18, 2019, www.pewsocialtrends.org/2018/01/09/women-and-men-in-stem-often-at-odds-over-workplace-equity; Wyer, Mary, "Intending to stay: Images of scientists, attitudes toward women, and gender as influences on persistence among science and engineering majors," *Journal of Women and Minorities in Science and Engineering* 9, no. 1 (2003): 1–16; Wyer, Mary, Jennifer Schneider, Sylvia Nassar-McMillan, and Maria Oliver-Hoyo, "Capturing stereotypes: Developing a scale to explore US college students' images of science and scientists," *International Journal of Gender, Science and Technology* 2, no. 3 (2010): 382–415.

23 Booksh, Karl S., and Lynnette D. Madsen, "Academic pipeline for scientists with disabilities," *MRS Bulletin* 43, no. 8 (2018): 625–632; Cech, Erin A., "LGBT professionals' workplace experiences in STEM-related federal agencies," paper presented at the 2015 ASEE Annual Conference & Exposition, Seattle, Washington, June 2015; Eaton, Asia A., Jessica F. Saunders, Ryan K. Jacobson, and Keon West, "How gender and race stereotypes impact the advancement of scholars in STEM: Professors' biased evaluations of physics and biology post-doctoral candidates," *Sex Roles* 82, no. 3–4 (2020): 127–141; Moran, Barbara, "Is science too straight? LGBTQ+ issues in STEM diversity," *The Brink: Pioneering Research from Boston University*, June 15, 2017, https://www.bu.edu/articles/2017/lgbt-issues-stem-diversity; O'Brien, Laurie T., Alison Blodorn, Glenn Adams, Donna M. Garcia, and Elliott Hammer, "Ethnic variation in gender-STEM stereotypes and STEM participation: An intersectional approach," *Cultural Diversity and Ethnic Minority Psychology* 21, no. 2 (2015): 169–180; Yoder, Jeremy B., and Allison Mattheis, "Queer in STEM: Workplace experiences reported in a national survey of LGBTQA individuals in science, technology, engineering, and mathematics careers," *Journal of Homosexuality* 63, no. 1 (2016): 1–27.

24 Belluz, Julia, and Brian Resnick, "Meat is unhealthy, meat is okay: Why science keeps overturning what we thought we knew," *Vox*, Oct. 4, 2019, www.vox.com/science-and-health/2019/10/4/20897383/nutrition-advice-science-meat-psychology-replication.

25 Brandt, Allan M., "Racism and research: The case of the Tuskegee Syphilis Study," *Hastings Center Report* (1978): 21–29; Flaherty, Dennis K., "The vaccine-autism connection: A public health crisis caused by unethical medical practices and fraudulent science," *Annals of Pharmacotherapy* 45, no. 10 (2011): 1302–1304.

26 Felt, Ulrike, Rayvon Fouché, Clark A. Miller, and Laurel Smith-Doerr, eds., *The handbook of science and technology studies*, MIT Press, 2017; Harding, Sandra, ed., *The postcolonial science and technology studies reader*, Duke University Press, 2011; Wyer, Mary, Mary Barbercheck, Donna Cookmeyer, Hatice Hatice Örün Öztürk, and Marta L. Wayne, eds., *Women, science and technology: A reader in feminist science studies*, Routledge, 2013.

27 Mooney, Chris, and Sheril Kirshenbaum, *Unscientific America: How scientific illiteracy threatens our future*, Basic Books, 2009, 4.

28 Nichols, Tom, "How does the public's view of science go so wrong?" *Scientific American*, Mar. 2, 2017, https://blogs.scientificamerican.com/guest-blog/how-does-the-public-rsquo-s-view-of-science-go-so-wrong.

29 National Science Board, "State of U.S. science."

30 Funk, Cary, and Brian Kennedy, "Public confidence in scientists has remained stable for decades," Pew Research Center, Mar. 22, 2019, www.pewresearch.org/fact-tank/2017/04/06/public-confidence-in-scientists-has-remained-stable-for-decades/; National Science Board, "State of U.S. science."

31 See the Appendix for details.

32 Pew Research Center, "Public and scientists' views on science and society," Pew Research Center, Jan. 29, 2015, www.pewinternet.org/2015/01/29/public-and-scientists-views-on-science-and-society.

33 Simis, Molly J., Haley Madden, Michael A. Cacciatore, and Sara K. Yeo, "The lure of rationality: Why does the deficit model persist in science communication?" *Public understanding of Science* 25, no. 4 (2016): 400–414.

34 Funk, Cary, "How much does science knowledge influence people's views on climate change and energy issues?" Pew Research Center, Mar. 22, 2017, www.pewresearch.org/fact-tank/2017/03/22/how-much-does-science-knowledge-influence-peoples-views-on-climate-change-and-energy-issues/.

35 National Academies of Sciences, Engineering, and Medicine, "Communicating science effectively: A research agenda," The National Academies, 2017.

36 Dudo, Anthony, Dominique Brossard, James Shanahan, Dietram A. Scheufele, Michael Morgan, and Nancy Signorielli, "Science on television in the 21st century: Recent trends in portrayals and their contributions to public attitudes toward science," *Communication Research* 38, no. 6 (2011): 754–777; Nisbet, Matthew C., Dietram A. Scheufele, James Shanahan, Patricia Moy, Dominique Brossard, and Bruce V. Lewenstein, "Knowledge, reservations, or promise? A media effects model for public perceptions of science and technology," *Communication Research* 29, no. 5 (2002): 584–608.

37 Brossard, Dominique, and Matthew C. Nisbet, "Deference to scientific authority among a low information public: Understanding US opinion on agricultural biotechnology," *International Journal of Public Opinion Research* 19, no. 1 (2007): 24–52; Ho, Shirley S., Dominique Brossard, and Dietram A. Scheufele, "Effects of value predispositions, mass media use, and knowledge on public attitudes toward embryonic stem cell research," *International Journal of Public Opinion Research* 20, no. 2 (2008): 171–192.

38 Brossard, Dominique, and Bruce V. Lewenstein, "A critical appraisal of models of public understanding of science," in *Communicating science: New agendas in communication*, eds. LeeAnn Kahlor and Patricia Stout, Routledge, 2010: 11–39; Irwin, Alan, and Brian Wynn, eds., *Misunderstanding science? The public reconstruction of science and technology*, Cambridge University Press, 2003.

39 Brossard and Lewenstein, "A critical appraisal"; Ley, Barbara L., *From pink to green: Disease prevention and the environmental breast cancer movement*, Rutgers University Press, 2009.

40 Editorial, "Science benefits from diversity," *Nature* 558, no. 5 (2018), www.nature.com/articles/d41586-018-05326-3; Gibbs, Kenneth, "Diversity in STEM: What it is and why it matters," *Scientific American*, Sept. 10, 2014, https://blogs.scientificamerican.com/voices/diversity-in-stem-what-it-is-and-why-it-matters; Hoy, Anne Q., "Leaders urged to support diversity in STEM ranks," American Association for the Advancement of Science, Jan. 9, 2017, www.aaas.org/news/leaders-urged-support-diversity-stem-ranks; Phillips, Katherine W., "How diversity works," *Scientific American* 311, no. 4 (2014): 42–47.

41 Helton, Mary, "100 Women: Where are the female Nobel Prize winners?" BBC News, Oct. 5, 2017, https://www.bbc.com/news/science-environment-41513261.

42 Holbrook, Jarita C., "Survival strategies for African American astronomers and astrophysicists," 2012, https://arxiv.org/pdf/1204.0247.pdf.

43 National Science Board, "State of U.S. science."

44 Cech, "LGBT professionals' workplace experiences"; National Science Foundation, "Women, minorities, and persons with disabilities in science and engineering," National Science Foundation/National Science Board, 2017, www.nsf.gov/statistics/2017/nsf17310/static/data/tab9-8.pdf.

45 Funk and Parker, "Women and men in STEM." For men, the percentages saying so were 28% and 7%, respectively.

46 Meanwhile, 13% of white respondents said the same.

47 Booksh and Madsen, "Academic pipeline"; Cech, "LGBT professionals' workplace experiences"; Moran, "Is science too straight?"; Yoder & Mattheis, "Queer in STEM."

48 Malcom, Shirley Mahaley, Paula Quick Hall, and Janet Welsh Brown, *The double bind: The price of being a minority woman in science,* American Association for the Advancement of Science, 1976.

49 Ong, Maria, Carol Wright, Lorelle Espinosa, and Gary Orfield, "Inside the double bind: A synthesis of empirical research on undergraduate and graduate women of color in science, technology, engineering, and mathematics," *Harvard Educational Review* 81, no. 2 (2011): 172–209.

50 Moran, "Is science too straight?"; Thurston, Linda P., Cindy Shuman, B. Jan Middendorf, and Cassandra Johnson, "Postsecondary STEM education for students with disabilities: Lessons learned from a decade of NSF funding," *Journal of Postsecondary Education and Disability* 30, no. 1 (2017): 49–60.

51 Moran, "Is science too straight?"

52 Desy, Elizabeth A., Scott A. Peterson, and Vicky Brockman, "Gender differences in science-related attitudes and interests among middle school and high school students," *Science Educator* 20, no. 2 (2011): 23–30; Gilmartin, Shannon K., Erika Li, and Pamela Aschbacher, "The relationship between interest in physical science/engineering, science class experiences, and family contexts: Variations by gender and race/ethnicity among secondary students," *Journal of Women and Minorities in Science and Engineering* 12, no. 2–3 (2006): 179–207; Gokhale, Anu A., Cara Rabe-Hemp, Lori Woeste, and Kenton Machina, "Gender differences in attitudes toward science and technology among majors," *Journal of Science Education and Technology* 24, no. 4 (2015): 509–516; Mallow, Jeffry, Helge Kastrup, Fred B. Bryant, Nelda Hislop, Rachel Shefner, and Maria Udo, "Science anxiety, science attitudes, and gender: Interviews from a binational study," *Journal of Science Education and Technology* 19, no. 4 (2010): 356–369; Riegle-Crumb, Catherine, Chelsea Moore, and Aida Ramos-Wada, "Who wants to have a career in science or math? Exploring adolescents' future aspirations by gender and race/ethnicity," *Science Education* 95, no. 3 (2011): 458–476.

53 Gerbner, George, "Science on television: How it affects public conceptions," *Issues in Science and Technology* 3, no. 3 (1987): 109–115.

54 Morgan, Michael, and James Shanahan, "The state of cultivation," *Journal of Broadcasting & Electronic Media* 54, no. 2 (2010): 337–355.

55 Iyengar, Shanto, and Donald R. Kinder, *News that matters: Television and American opinion,* University of Chicago Press, 2010; Scheufele, Dietram A., and David Tewksbury, "Framing, agenda setting, and priming: The evolution of three media effects models," *Journal of Communication* 57, no. 1 (2007): 9–20.

56 Myrick, Jessica Gall, and Suzannah D. Evans, "Do PSAs take a bite out of shark week? The effects of juxtaposing environmental messages with violent images of shark attacks," *Science Communication* 36, no. 5 (2014): 544–569.

57 Brewer, Paul R., and Barbara L. Ley, "Media use and public perceptions of DNA evidence," *Science Communication* 32, no. 1 (2010): 93–117; Pettey, Gary R., and Cheryl Campanella Bracken, "The *CSI* effect: Scientists and priming on prime time television," in *Common sense: Intelligence as presented on television,* ed. Lisa Holderman, Lexington Books, 2008: 233–47.

58 Gamson, William A., *Talking politics*, Cambridge University Press, 1992; Goffman, Erving, *Frame analysis: An essay on the organization of experience*, Harvard University Press, 1974; Tversky, Amos, and Daniel Kahneman, "The framing of decisions and the psychology of choice," *Science* 211, no. 4481 (1981): 453–458.

59 Druckman, James N., "What's it all about? Framing in political science," in *Perspectives on framing*, ed. Gideon Keren, Psychology Press, 2011, 279–302; Gamson, William A., and Andre Modigliani, "Media discourse and public opinion on nuclear power: A constructionist approach," *American Journal of Sociology* 95, no. 1 (1989): 1–37; Iyengar, Shanto, *Is anyone responsible? How television frames political issues*, University of Chicago Press, 1994.

60 Nisbet, Matthew C., and Mike Huge, "Attention cycles and frames in the plant biotechnology debate: Managing power and participation through the press/policy connection," *Harvard International Journal of Press/Politics* 11, no. 2 (2006): 3–40.

61 Brewer, Paul R., "The trappings of science: Media messages, scientific authority, and beliefs about paranormal investigators," *Science Communication* 35, no. 3 (2013): 311–333; Brewer, Paul R., and Jessica McKnight, "'A statistically representative climate change debate': Satirical television news, scientific consensus, and public perceptions of global warming," *Atlantic Journal of Communication* 25, no. 3 (2017): 166–180.

62 Bandura, Albert, "Human agency in social cognitive theory," *American Psychologist* 44, no. 9 (1989): 1175–1184; Bandura, Albert, "Social cognitive theory of mass communication," *Media Psychology* 3, no. 3 (2001): 265–299; Long, Marilee, and Jocelyn Steinke, "The thrill of everyday science: Images of science and scientists on children's educational science programmes in the United States," *Public Understanding of Science* 5, no. 2 (1996): 101–120.

63 Ryan, Lisa, and Jocelyn Steinke, "'I want to be like...': Middle school students' identification with scientists on television," *Science Scope* 34, no. 1 (2010): 44–49; Steinke, Jocelyn, Brooks Applegate, Maria Lapinski, Lisa Ryan, and Marilee Long, "Gender differences in adolescents' wishful identification with scientist characters on television," *Science Communication* 34, no. 2 (2012): 163–199.

64 O'Keeffe, Moira, "Lieutenant Uhura and the drench hypothesis: Diversity and the representation of STEM careers," *International Journal of Gender, Science and Technology* 5, no. 1 (2013): 4–24; Steinke, Jocelyn, Maria Lapinski, Marilee Long, Catherine Van Der Maas, Lisa Ryan, and Brooks Applegate, "Seeing oneself as a scientist: Media influences and adolescent girls' science career possible selves," *Journal of Women and Minorities in Science and Engineering* 15, no. 4 (2009): 270–301.

65 Dudo et al., "Science on television."

66 Nisbet et al., "Knowledge, reservations, or promise?"

67 Feldman, Lauren, Edward W. Maibach, Connie Roser-Renouf, and Anthony Leiserowitz, "Climate on cable: The nature and impact of global warming coverage on Fox News, CNN, and MSNBC," *International Journal of Press/Politics* 17, no. 1 (2012): 3–31.

68 Brewer, Paul R., and Jessica McKnight, "Climate as comedy: The effects of satirical television news on climate change perceptions," *Science Communication* 37, no. 5 (2015): 635–657.

69 Krippendorff, Klaus, *The content analysis reader*, Sage, 2009.

70 Asher, Herb, *Polling and the public: What every citizen should know*, CQ Press, 2016.

71 Funk et al., "Science news."

72 Eveland, William P., Jr., and Douglas M. McLeod, "The effect of social desirability on perceived media impact: Implications for third-person perceptions," *International Journal of Public Opinion Research* 11, no. 4 (1999): 315–333.

73 Hastie, Reid, and Bernadette Park, "The relationship between memory and judgment depends on whether the judgment task is memory-based or on-line," *Psychological Review* 93, no. 3 (1986): 258–268.

74 Iyengar and Kinder, *News that matters*.

2

MOVIE SCIENCE

Henry Wu: You are acting like we are engaged in some kind of mad science. But we are doing what we have done from the beginning. Nothing in Jurassic World is natural. We have always filled gaps in the genome with the DNA of other animals. And, if their genetic code was pure, many of them would look quite different. But you didn't ask for reality. You asked for more teeth.

—*Jurassic World* (2015)

Mark Watney: In the face of overwhelming odds, I'm left with only one option: I'm gonna have to science the shit out of this.

—*The Martian* (2015)

In 2015, moviegoers flocked to two science-themed Hollywood films. One was *Jurassic World*, a reboot of a popular franchise that began with 1993's *Jurassic Park*. The original film revolves around an ill-fated theme park featuring dinosaurs— including velociraptors and a Tyrannosaurus rex—recreated through cloning. *Jurassic World* introduces a new theme park and a genetically engineered hybrid dinosaur that can camouflage itself. The creature escapes to wreak havoc, rein- forcing the two overarching messages of the series: that tampering with nature out of hubris (as in the original *Jurassic Park*) or corporate greed (as in *Jurassic World*) is dangerous, and that watching dinosaurs on the rampage is fun for the whole family. The reboot was the second-highest grossing movie of the year, earning more than $650 million at the box office.[1] Inevitably, a sequel followed in 2018.

DOI: 10.4324/9781003190721-2

Whereas *Jurassic World* focuses on using computer-generated imagery (CGI) to present the spectacle of dinosaurs attacking humans and one another, *The Martian* tells the story of Mark Watney, an astronaut marooned alone on Mars who uses his scientific knowledge to survive for more than a year and a half until rescuers arrive. His training as a botanist helps him farm the planet's soil with water produced from rocket fuel and fertilizer produced from his own feces. *The Martian* not only earned the eighth-highest box office take for the year (more than $200 million) but also scored an Academy Award nomination for Best Picture as well as a Best Actor nomination for Matt Damon (who portrays Watney).

The depictions of science in *Jurassic World* and *The Martian* reflect film traditions that date back more than a century to the very first science fiction movie, *A Trip to the Moon* (*Le Voyage dans la Lune*). Like *The Martian*, this 1902 French film features a space expedition gone awry. Like *Jurassic World*, it also presents creatures—here, the Selenites who inhabit the Moon—that dramatize both the wonders and the dangers of scientific inquiry.

The first wave of science fiction movies started a quarter century later with *Metropolis*, a 1927 German film that helped to popularize such tropes as futuristic dystopian societies, human-like robots, and mad scientists who play god. In 1931, the US film industry elevated the genre to new heights when it brought one of the most iconic scientist characters of all time to the silver screen: Dr. Frankenstein, whose shouts of "It's alive!" from his electrode-filled laboratory epitomize Hollywood mad science. When the Universal Pictures adaptation of Mary Shelley's novel became a hit, the studio and its rivals unleashed an army of evil scientists in films such as *Island of Lost Souls* (1932), *The Invisible Man* (1933), and *The Bride of Frankenstein* (1935).

During the Cold War, Hollywood shifted its focus to new scientific themes. Some movies from this era depict epic disasters, such as the destruction of the Earth by a "rogue star" in *When Worlds Collide* (1951). Others feature atomic-age monsters, such as the bomb-revived monster in *The Beast from 20,000 Fathoms* (1953), the mutated giant ants of *Them!* (1954), and the famous radioactive Japanese import *Godzilla* (1954). Still others—foremost among them *The War of the Worlds* (1953) and *Invasion of the Body Snatchers* (1956)—portray alien invasions partly inspired by media reports of UFOs (unidentified flying objects).

The extraterrestrial invasion genre has survived to this day, with recent entries including *Battleship* (2012), *Pacific Rim* (2013), and *Edge of Tomorrow* (2014). Likewise, the disaster genre has lived on in films about volcanoes (*Dante's Peak* and *Volcano*, both 1997), asteroids (*Armageddon*, 1998), comets (*Deep Impact*, 1998), climate change (*The Day After Tomorrow*, 2004), and even "mutating neutrinos" (*2012*, 2009). Scientific themes in Hollywood movies from the past few decades also include artificial intelligence (*A.I.: Artificial Intelligence*, 2001; *I, Robot*, 2004), communication with alien life (*Contact*, 1997; *Arrival*, 2016), genetic engineering (*Gattaca*, 1997; *Rampage*, 2018), and pandemics (*Contagion*, 2011).

Some of these movies have bombed at the box office, but others have been hits. Indeed, the top-grossing films of the 2010s include a number of science-themed movies besides *Jurassic World* and *The Martian*. *Gravity*, which focuses on a biomedical engineer stranded in space, was the eighth-highest grossing film of 2013 and won Academy Award nominations for Best Picture and Best Actress. The following year, *Interstellar* (2014) placed tenth at the box office with its story about astronauts searching for a habitable planet to replace a devastated Earth. In 2016, *Hidden Figures* earned more than $150 million and a Best Picture nomination for its portrayal of three Black women working as mathematicians at NASA during the 1960s. Recent blockbusters have also featured scientists as action heroes (such as physicist Bruce Banner from 2012's *The Avengers*) or villains (such as chemist Isabela Maru from 2017's *Wonder Woman*).

Many moviegoers no doubt view these portrayals as mere entertainment. Over the years, however, some scientists have criticized the film industry for presenting inaccurate science and depicting scientists themselves as strange or even menacing. Chris Mooney and Sheril Kirshenbaum capture this perspective in their 2009 book *Unscientific America*:

> There's a long litany of complaints: Too many stereotypical nerdy scientists ... and too few positive role models. Too many mad scientists trying to play God ... And too many simply ridiculous "scientific" premises.[2]

Those who raise such criticisms worry that moviegoers will come away with a distorted image of science that fuels misperceptions, damages public support for scientific research, and dampens young people's interest in the profession. Another common concern is that Hollywood perpetuates the stereotype of the scientist as a white, heterosexual, able-bodied man, thereby discouraging people who don't match this stereotype from pursuing scientific education and careers.[3] At the same time, some scientists see the film industry's fascination with science as an opportunity to publicize important findings and build popular enthusiasm for new research.

Misleading or not, Hollywood portrayals of science use an array of techniques to create a sense of plausibility and realism. With this in mind, we explore how such portrayals influence the movie-watching public. Do they shape beliefs about—and support for—endeavors such as space exploration and the revival of extinct species through genetic engineering? Do film depictions of scientists sway perceptions of their real-world counterparts, perhaps even inspiring audience members to emulate them as role models? And do recent movies reinforce demographic stereotypes of scientists or offer a more inclusive portrait of the profession, one that could help foster greater diversity in STEM?

Scientific Accuracy versus Scientific Plausibility in Hollywood Films

When scientists themselves review movies, they often concentrate on dissecting what facts the filmmakers got right or wrong. For example, physicist Sidney Perkowitz's 2007 book *Hollywood Science* examines whether movie portrayals have a "real basis" and could "actually happen."[4] The author bestows a "Golden Eagle" award on *Jurassic Park* for "showing viewers the immense possibilities of modern genetics," and he gives *The Day After Tomorrow* a "Special Award" for raising an alarm about the real-world risks from climate change.[5] Perkowitz also awards "Golden Turkeys" for particularly egregious movie portrayals of science, with one going to *The Core* for its absurd depiction of scientists drilling to the center of the Earth so they can start the planet spinning again with a hydrogen bomb.

In recent years, Twitter has become a particularly popular platform for film commentary by scientists. For instance, one paleontologist, Darren Naish, tweeted his frustration with the absence of feathers on the dinosaurs in *Jurassic World*: "Basic message of #JurassicWorld is 'Screw you, science, we don't need your stinking feathers!! This is alt-future-1993!!' #dinosaurs."[6] More plaintively, Morgan Jackson tweeted, "Dear #JurassicWorld, Entomologists are real. We're actually quite nice. Please ask us why mosquitos/= crane flies."[7] Meanwhile, *The Martian* elicited a glowing endorsement from astrophysicist Neil deGrasse Tyson. "The @MartianMovie—where you learn all the ways that being Scientifically Literate can save your life," he tweeted. "[T]hey got crucial science right, while enhancing the story by fictionalizing the science that remained."[8]

Yet accuracy is seldom Hollywood's primary goal in depicting science. Instead, filmmakers tend to care more about telling plausible stories that the audience will enjoy. To this end, they strive to create a credible-seeming "look" and "sound" of science that matches moviegoers' expectations and thereby helps maintain the "willing suspension of disbelief."[9] Communication scholar David Kirby argues that such *perceptual realism*, rather than factual realism, is the key to understanding how audience members respond to Hollywood science.[10] For example, they may judge the plausibility of *Jurassic World* less on whether its portrayals of T-rexes and velociraptors reflect authentic science than on whether its special effects, sound design, dialogue, plot, and other cinematic elements make these depictions *feel* real in the context of the movie's world and viewers' own mental images of dinosaurs.

Filmmakers use a variety of techniques to create perceptually realistic movie science. In terms of visuals, production teams often cater to—and in doing so, may reinforce—cultural stereotypes by dressing scientist characters in white lab coats, equipping them with notebooks, and surrounding them with equation-filled chalkboards, liquid-filled test tubes, or elaborate computer screen displays.

As a case in point, *Jurassic World* presents geneticist Henry Wu wearing a lab coat (albeit a stylish gray one over a black turtleneck) in a laboratory decked with microscopes, screens, and pipettes. Screenwriters also give scientists dialogue filled with impressive-sounding technical jargon to replicate the language of science. For example, astronaut/botanist Mark Watney from *The Martian* measures time in "sols" instead of days and uses terms such as "the Hab" (the Mars Lander Habitat), "the MDV" (Mars Descent Vehicle), and "EVA suits" (Extra-Vehicular Activity suits).

As Kirby describes in his 2011 book *Lab Coats in Hollywood: Science, Scientists, and Cinema*, filmmakers often hire scientists as consultants to help enhance the plausibility of their movies. For example, astronomer Josh Colwell introduced the makers of *Deep Impact* to the notion of comet "outgassing" (they initially thought he was joking but ultimately incorporated the phenomenon into the plot) and suggested substituting Titan missiles for Ariane missiles when actor Morgan Freeman had difficulty pronouncing the latter name.[11] Similarly, a real-life mathematician served as actor Russell Crowe's "hand double" in the biographic film *A Beautiful Mind* (2001) so that the main character would appear to be writing equations with a "natural flow."[12] Still, filmmakers sometimes discard the advice of their consultants in favor of playing to popular misconceptions of science. In the case of *The Nutty Professor* (1996), one consultant Kirby interviewed tried to create a realistic-looking biology lab only for the director to demand "red and green solutions bubbling and boiling."[13]

Hollywood's practice of hiring science consultants has continued since Kirby published his book. Take *The Martian* director Ridley Scott, who went straight to NASA for advice. As Jim Green, the director of the agency's Planetary Science Division, explained, Scott "wanted to know what NASA was doing in developing habitats, and what our vehicles looked like, so that he could model what the *look and feel* of the movie was all about based on what we're really doing" (emphasis added).[14] Green responded by inviting the film's production team to tour the agency's laboratory and sharing mockups of NASA's plans for visiting the planet. Despite a few unrealistic touches, such as a dust storm in the thin Martian atmosphere powerful enough to knock over large objects, Green gave the movie high marks for its plausibility.

For their part, the makers of *Jurassic World* turned to Jack Horner, the paleontologist who consulted for the original *Jurassic Park* films and helped inspire the first movie's hero, Dr. Alan Grant.[15] According to Horner, the makers of *Jurassic World* "wanted the dinosaurs to be as accurate as possible. But once they are *accurate looking*, they become the actors and they obviously do things that real animals wouldn't do—like chasing people all over the place and breaking into buildings just to eat a person" (again, emphasis added).[16] In interviews, he acknowledged some of the film's inaccuracies, noting that obtaining dinosaur DNA from mosquitos in amber is unrealistic and that dinosaurs probably neither roared nor growled (instead, they may have sung like birds).[17] However, Horner explained away the absence of dinosaur feathers in the movie—a pet peeve among many other paleontologists—by noting that the *Jurassic World* scientists genetically engineered their dinosaurs.

Many of Kirby's interviewees consulted on movies not only to help Hollywood depict plausible science but also to highlight specific issues. For example, Josh Colwell saw his work for the comet-strikes-Earth film *Deep Impact* as helping to publicize the real-world risks posed by "Near Earth Objects," while climatologist Michael Molitor viewed *The Day After Tomorrow*'s scientific scenario as an opportunity to warn the public of the dangers from climate change. Meanwhile, geophysicist J. Marvin Herndon "actively sought out" the makers of *The Core* to publicize his own (widely rejected) theory that the earth has a natural nuclear reactor at its center that will burn out with catastrophic consequences.[18]

Similar concerns about fostering public engagement and shaping public expectations helped motivate both Green's work on *The Martian* and Harmon's work on the *Jurassic Park* franchise. "We are funded by the public," Green explained to one journalist. "It's important to us that they see we're making major progress in planetary science."[19] Harmon also suggested that nitpicking the scientific accuracy of Hollywood films is ultimately misplaced. "Every time a *Jurassic Park* movie comes out," he argued, "it raises the public's awareness about dinosaurs ... If it were a documentary ... it wouldn't get near as many people to it."[20]

Movie Effects on Audience Perceptions of Science

Green and Harmon were onto something: when filmmakers succeed in creating plausible depictions of science, such portrayals can shape audience members' impressions. In some cases, watching inaccurate movie science leads to misperceptions—as with *The Core*'s wildly erroneous geology. When a team of researchers led by Michael Barnett divided students at a middle school into two groups and showed one group this movie, the students who saw it came away with more distorted understandings of earth science.[21] Instead of citing the factual information they'd learned during eight weeks of instruction, the movie-watchers tended to mention *The Core*'s more fanciful depictions of geology. In keeping with Kirby's argument about the importance of perceptual realism, these young viewers saw the movie as plausible because its hero, a geophysics professor, seemed credible to them—particularly after an early scene in which he correctly describes the earth's structure. By grounding the premise of *The Core* in believable details, its makers created footholds for scientific misinformation.

In other cases, flawed but perceptually realistic film portrayals can promote acceptance of mainstream science. Take *The Day After Tomorrow*, a box office hit that scientists criticized for its inaccurate depictions of rapid global freezing but also praised for its warning about climate change. To study this movie's impact on public perceptions, Anthony Leiserowitz surveyed the US public three weeks after its release.[22] Consistent with *The Day After Tomorrow*'s scientific scenario, respondents who had seen it were especially likely to perceive a new Ice Age as possible. At the same time, they reported greater concern about global warming and greater perception of risks from it, including increasingly powerful storms and flooding of major cities. Furthermore, the respondents who had seen the movie were more likely to say they would take future steps to reduce global warming, including changing their purchasing habits, joining or donating to groups, and talking to friends, family, and politicians about the issue.

Our own investigation of the links between movie watching and perceptions of science focused on two topics featured in recent films: sending astronauts to Mars (as in *The Martian*) and reviving extinct species (as in *Jurassic World*). The first of these is a key goal of NASA, which has developed a three-stage plan for sending humans to the planet by the 2030s.[23] "Mars is the next tangible frontier for human exploration, and it's an achievable goal," proclaims the agency's website, "There are challenges to pioneering Mars, but we know they are solvable. We are well on our way to getting there, landing there, and living there."[24] Several other organizations, among them entrepreneur Elon Musk's aerospace corporation SpaceX, have announced their own ambitions of sending humans to the Red Planet.[25] To be sure, some experts—including Neil deGrasse Tyson—doubt that humans will actually travel to Mars, largely due to the massive expenses involved.[26] Funding issues aside, however, most experts agree that such a mission is at least feasible in the near future.[27]

The notion of reviving long-extinct species through genetic science has proven more controversial within the scientific community. The idea has its advocates, including *Jurassic World* consultant Jack Horner, who proposed a real-life plan for doing so in his 2009 book with James Gorman, *How to Build a Dinosaur: Why Extinction Doesn't Have to Be Forever*.[28] Similarly, biologist and science journalist Helen Pilcher advocates reviving dinosaurs and other species in her 2017 book, *Bring Back the King: The New Science of De-extinction*.[29] Yet any effort to recreate a Tyrannosaurus rex would face a currently insurmountable obstacle: the absence of dinosaur DNA, which naturally decays over time (*Jurassic Park*'s depiction of retrieving viable samples from ancient mosquito blood notwithstanding). Even recreating the woolly mammoth, which once co-existed with humanity, involves daunting scientific challenges—as biologist Beth Shapiro describes in her book, *How to Clone a Woolly Mammoth*.[30] Some scientists have also raised ethical concerns regarding de-extinction, much as fictional mathematician Ian Malcolm does in the original *Jurassic Park*.[31] Nevertheless, a team led by Harvard geneticist George M. Church is presently working on a project to engineer a mammoth-elephant hybrid, or "mammophant," capable of living in cold climates.[32]

To capture how movie-watching shapes beliefs about sending astronauts to Mars and reviving extinct species, we included four questions on a nationally representative survey of 1,000 US residents conducted in October 2016.[33] Two questions asked respondents whether they *favored or opposed* "the United States sending astronauts to explore Mars" and "scientists bringing an extinct species back to life, such as the woolly mammoth." Fully 77% of the public supported the first endeavor, whereas only 39% supported the second. The other pair of questions asked respondents how *likely* they thought it was that each would happen "before the year 2050." An overwhelming 82% of the public expected the United States to send humans to Mars by 2050, compared to 43% who expected scientists to revive an extinct species such as the woolly mammoth. These perceptions dovetail with mainstream views within the scientific community: most experts see a Mars mission as both desirable and attainable, whereas de-extinction is more controversial among experts in terms of both its feasibility and wisdom.

Did movie-watchers and non-watchers differ in their perceptions of each topic? To answer this question, we included another set of questions on the survey asking whether respondents had seen four science-themed movies: *Jurassic World* (53% had done so), *The Martian* (39%), *Gravity* (42%), and *Interstellar* (28%). Respondents who'd seen any one of these movies tended to have seen the others, too, so we created a combined measure of how many of the four movies each respondent had watched. Almost half the sample (46%) had seen at least two of the movies, reflecting the broad reach of science-themed Hollywood blockbusters.

Support for a Mars mission differed depending on how many of the four movies respondents had watched, as did support for de-extinction (Figure 2.1). Of the respondents who had watched one or none of the movies, 73% favored sending astronauts to Mars and 29% favored bringing extinct species back to life. By comparison, support among those who had watched two or more of the films was 9 points greater for a Mars mission (82%) and 13 points greater for de-extinction (42%).

We found similar patterns in our respondents' predictions of what would happen by 2050. Of those who had seen one or none of the four movies, 77% expected the United States to send astronauts to Mars and 37% expected scientists to bring an extinct species back to life. Respondents who had seen two or more of the movies had greater expectations: they were 11 points more likely to predict that a Mars mission (88%) and de-extinction would happen (48%).

All the relationships we found here remained statistically significant even after taking into account the roles of demographic factors and other forms of media use. So, what explains these patterns? The link between science-movie watching and support for sending astronauts to Mars makes sense given the

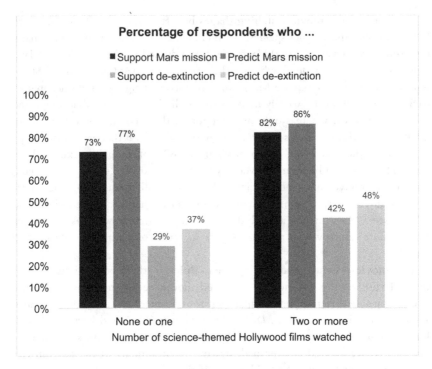

Percentage of respondents who ...

■ Support Mars mission ■ Predict Mars mission
■ Support de-extinction ■ Predict de-extinction

FIGURE 2.1   Opinions and expectations about scientific endeavors, by movie watching (Cooperative Congressional Election Survey, 2016)

pro-space exploration messages of not only *The Martian* but also *Gravity* and *Interstellar*. Moreover, the plausible portrayals of human-crewed space missions in these movies may have encouraged watchers to perceive a voyage to Mars as feasible.

By the same logic, the realistic-seeming look and language of *Jurassic World*'s science may have fostered perceptions that de-extinction is possible or even likely. The relationship between movie watching and *support* for reviving extinct species is more surprising at first glance, particularly in light of *Jurassic World*'s cautionary tale about the dangers and ethical problems of doing so. In part, this pattern may reflect the movie's other central message: that reviving extinct specific is *cool*. The characters in the film react to its hybrid dinosaurs with awe and wonder (until the chaos begins), and the filmmakers clearly intended for audience members to respond the same way. More broadly, all four movies convey the message that scientific inquiry can be dangerous but can also open the door to amazing new discoveries.

By creating plausible depictions of space travel and dinosaurs, the makers of *The Martian* and *Jurassic World* helped persuade movie-goers to buy tickets, suspend their disbelief, and enjoy the show. In the process, the filmmakers may have swayed perceptions of real-world science in ways that promoted the goals of their consultants. Consistent with NASA's and Jack Horner's ambitions, Americans who'd seen science-themed films were particularly likely to support a Mars mission and de-extinction, as well as to anticipate success in both endeavors.

But what about perceptions of scientists themselves? A look at Hollywood's portrayals of the scientific profession over the past century reveals that they've reflected—and perhaps also shaped—audience expectations of what scientists are really like.

Hollywood Scientists: Mad, Odd, or Heroic?

During the early years of the film industry, the typical movie scientist was a god-playing madman (but definitely not a mad *woman*).[34] The title character of *Frankenstein* is the archetypal 1930s Hollywood scientist: he uses electricity to create a misunderstood monster that goes berserk when humans reject it. Likewise, Dr. Moreau from *Island of Lost Souls* engineers human-animal hybrids that ultimately turn on him. In *The Invisible Man*, Dr. Griffin tries to use his invisibility formula to take over the world. Dr. Gogol of *Mad Love* (1935) grafts the hands of a murderer onto an injured pianist, while Dr. Ruhk of *The Invisible Ray* (1936) undertakes a killing spree after radium from outer space gives him a deadly touch. And so on.

Although Hollywood's portrayals of characters such as Dr. Frankenstein were exaggerated, they reflected audience members' anxieties about the potential for science to change their way of life, threaten their values, and pose new dangers.[35] Name-checking a famous tale of mad science, film scholar David Skal argues that moviegoers of the era held a "Jekyll and Hyde" stance toward science: laypeople were culturally and economically dependent on its contributions, but many worried about its influence.[36] "The real scientist professes to be motivated by the pursuit of rational, objective knowledge," Skal concludes. "His mad counterpart, like the fool in a king's court, has the license to speak more plainly of his motivations, which can constitute the spectrum of human venality, with a special emphasis on power."[37]

Over time, the character of the mad scientist became so familiar that films began to parody it. For example, the title character in the 1964 dark comedy *Dr. Strangelove* is a doomsday-device-building German nuclear expert modeled on real-life NASA rocket scientist Wernher von Braun. Meanwhile, Dr. Frederick Frankenstein (pronounced Fronk-in-STEEN, as he repeatedly emphasizes) from the 1974 horror comedy *Young Frankenstein* is the grandson of Hollywood's archetypal mad scientist; like his ancestor, he tries but fails to resist the lure of forbidden science. Dr. Frank N. Furter from the musical *The Rocky Horror Picture Show* (1975) is a similarly affectionate parody of the mad scientist who plays god in a laboratory—in this case, by creating, not a monster like Dr. Frankenstein's, but a "perfect man."

Yet the classic, non-parody incarnations of this trope have become increasingly rare specimens, prompting Rosylnn D. Haynes to ask, "Whatever happened to the 'mad, bad' scientist?"[38] She argues that decline of such characters reflects "decreasing ignorance and fear of science and increasing acceptance of scientists as professional members of society."[39] She also points to growing public concerns about environmental issues, where scientists can offer potential solutions; the increasing inclusion of women scientist characters, seldom ever portrayed as "mad"; and the rise of new Hollywood villains to take on the role of the frightening "other."[40] Apart from throwbacks such as Dr. Heiter from the 2009 shock horror movie *The Human Centipede*, sinister scientists have given way to terrorists, dictators, and corporate overlords.

In place of the vanishing mad scientist who seeks to take over the world or play god, Hollywood has introduced several other archetypes for the profession. One is the scientist as an eccentric nerd or geek.[41] Early cases include Professor Brainard from 1961's *The Absent-Minded Professor*, who invents an amazing substance he calls "flubber," and Professor Kelp from 1964's *The Nutty Professor*, who invents a serum that transforms him into a charming but mean-spirited hipster (Hollywood remade both movies in the 1990s). A more recent instance is the wild-eyed, disheveled Dr. Orun from *Independence Day* (1996).[42] Still, perhaps the most famous example of this archetype is Doc Brown, the inventor of the time-traveling DeLorean car from the *Back to the Future* film franchise (1985–1990). As portrayed by Christopher Lloyd, he's a blend of the "mad" and "nerd" stereotypes who wears a white lab coat, has wild white hair, and peppers his speech with references to "gigawatts" and "the flux capacitor" along with shouts of "Great Scott!" Such portrayals mirror enduring and widely held, though far from universal, public stereotypes of scientists as odd and peculiar.[43]

Another archetype in recent films is the scientist as a pawn of the government, industry, or some other powerful force.[44] Examples of this category include the geneticists from the original *Jurassic Park* (1993) and the cloning-themed action movie *The 6th Day* (2000), along with the gullible Dr. Powell from *Bumblebee* (2018).[45] In some cases, these scientists fall prey to professional ambition and greed as they work in new, relatively unregulated fields such as cloning and biotechnology.[46] Dr. Wu from *Jurassic World* is a typical case: he creates the dangerous hybrid *Indominus rex* after the theme park's executives demand new, scarier dinosaurs to entice a public grown jaded by conventional dinosaurs. With its focus on scientists as puppets of corrupt leaders or corporations, this archetype reflects a deepening public cynicism toward institutions such as government and big business.[47]

Over the past few decades, a different Hollywood archetype has become increasingly common: the heroic scientist.[48] The 1990s saw a wave of courageous scientists, including paleontologist Alan Grant from *Jurassic Park*, astronomer Ellie Arroway from *Contact*, and volcanologist Harry Dalton from *Dante's Peak*. The trend continued in the next decade with scientist heroes such as geophysicist Josh Keyes from *The Core* and paleoclimatologist Jack Hall from *The Day After Tomorrow*. Furthermore, it has extended to the 2010s with the likes of biomedical engineer Ryan Stone from *Gravity*, biologist Amelia Brand from *Interstellar*, and primatologist Davis Okoye from *Rampage*. In keeping with the public's growing tendency to see scientists as people dedicated to the good of humanity, these characters use their knowledge for noble ends that range from making contact with alien life (*Arrival*) to saving the Earth (*The Core*).[49] On top of that, they engage in action movie derring-do by rescuing children from the dinosaurs (*Jurassic Park*), lava (*Dante's Peak*), or cold air (*The Day After Tomorrow*). This being Hollywood, they tend to possess movie-star good looks, too.

Mark Watney from *The Martian* epitomizes the heroic scientist of contemporary Hollywood films. Confronted with months of danger and isolation, he responds with ingenuity, determination, and humorous bravado. At one point, he says, "Mars will come to fear my botany powers." At another, he jokingly tells NASA, "I don't mean to sound arrogant or anything, but I am the greatest botanist on this planet." *The Martian* portrays the other astronauts on his mission as resourceful and willing to risk their own lives for his. Likewise, the movie depicts the scientists and engineers at NASA—as well as the Chinese National Space Administration, which lends assistance—as dedicated to their jobs and the safety of the astronauts on the mission.

A few elements of the nerd archetype do appear in *The Martian*. For example, a scene depicts a group of NASA leaders as having a geeky sense of humor when they dub a secret meeting "Project Elrond" (a reference to *Lord of the Rings*). Additionally, one NASA astrodynamacist comes across as brilliant but so eccentric that he fails to recognize the director of his own agency. These portrayals are affectionate, however, and the dominant image of scientists in the movie is overwhelmingly positive. Characters may sometimes disagree with one another or make decisions that leave Watney in danger—as when the NASA director initially vetoes a rescue attempt—but they do so for rational and justifiable reasons.

With the rise of the movie hero scientist, some observers even worry about Hollywood falling into "science worship." For example, *Smithsonian* editor Rachel Gross argues that *The Martian* essentially depicts science as "magic" and scientists as "wizards." "Consider a key scene in the movie, when a plan to rescue Watney fails," she writes. "But hark! Rich Purnell ... a disheveled NASA astrophysicist, swiftly comes up with another plan—one involving a slingshot maneuver and some complex equations. The film doesn't explain how Purnell came to this conclusion. It doesn't explain why anybody should believe him. He just downs a carafe of coffee and it comes to him in a stroke of divine inspiration."[50] Gross warns that portraying "science as magic ... encourages a blind faith in scientists as masters of the universe." Her criticisms echo a broader theme in the field of science and technology studies: the tendency of our society to cast science a pathway to truth lit by flashes of individual brilliance rather than a lengthy, incremental, and collaborative process.[51]

Just as the film industry's plausible-seeming portrayals of space missions and de-extinction can shape public perceptions, so may its depictions of scientists. Everyone involved in making movies such as *The Martian*—from the directors and screenwriters to the science consultants to the actors—strives to create believable characters. By portraying scientists as heroes or oddballs, they could help reinforce these popular archetypes in audience members' minds.

Movie-Watching and Perceptions of Scientists

To explore whether moviegoers and non-moviegoers differ in the ways they perceive scientists, we included two questions on our October 2016 survey of the US public. The first asked respondents how much they agreed or disagreed that "scientists are dedicated people who work for the good of humanity." The second asked them how much they agreed or disagreed that "scientists tend to be odd and peculiar people." In effect, we measured whether members of the public see real-world scientists as being like Mark Watney from *The Martian*, Doc Brown from *Back to the Future*, or both.

As it happens, watching Hollywood science movies went hand in hand with perceiving scientists as good but not with perceiving them as strange. Of the respondents who had seen one or none of the four movies we asked about (*The Martian, Jurassic World, Gravity,* and *Interstellar*), 21% strongly agreed that scientists are dedicated people who work for good (Figure 2.2). By contrast, 31% of those who had seen two or more of the movies strongly agreed with this statement. This ten-point difference was statistically significant even after controlling for demographic factors and other forms of media use. Furthermore, it fits with the recent trend of Hollywood movies portraying scientists as heroic rather than evil.

Meanwhile, movie-watchers and non-watchers reported almost identical views on whether scientists are odd and peculiar (Figure 2.3). Of the respondents who had seen one or none of the four movies, 6% strongly agreed with this statement and 42% agreed with it. Of those who had seen two or more of the movies, the figures were 9% and 37%, respectively. Nor were the differences between watchers and non-watchers statistically meaningful here after we took into account the role of other factors. Watching contemporary Hollywood movies doesn't seem to fuel perceptions of scientists as weird or antisocial, let alone mad or villainous.

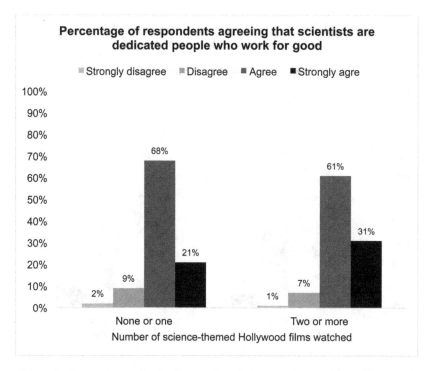

Percentage of respondents agreeing that scientists are dedicated people who work for good

FIGURE 2.2 Perceptions of scientists as good, by movie watching (Cooperative Congressional Election Survey, 2016)

Viewed in this light, complaints that the film industry depicts "too many stereotypical nerdy scientists" and "too few positive role models" have it backwards.[52] If anything, recent movies foster an image of scientists as admirable role models to emulate—or even as superhuman. Yet Hollywood's power to inspire future scientists raises another concern: whether the film industry gives everyone the same opportunity to be a scientist on the big screen.

The Demographics of Hollywood Science

In casting scientists as madmen, oddballs, or, increasingly, heroes, Hollywood also stereotypes them in ways that may convey signals about who can and cannot work in science. To start with, Hollywood has a long history of portraying scientists as men. Looking at scientist characters in 222 movies from the 1930s to the end of the 20th century, Peter Weingart and his colleagues found that only one in five were women—a pattern that mirrored the gender gap in real-world

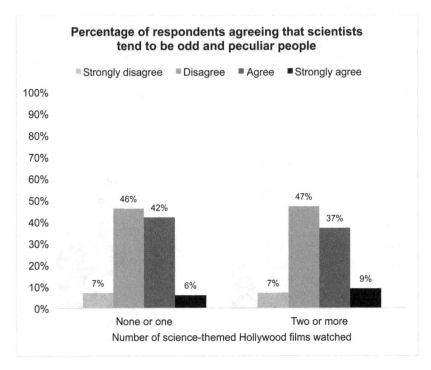

FIGURE 2.3 Perceptions of scientists as odd and peculiar, by movie watching (Cooperative Congressional Election Survey, 2016)

science, and maybe audience expectations as well.[53] In her analysis of 60 films from 1929 to 2003, Eva Flicker also found that the depictions of such characters relied on a different set of archetypes than the ones typically used for men. Instead of maniacs, nerds, or heroes, women scientists in movies tended to be "old maids" who dedicated themselves to science above all else; "male women" who took on stereotypically masculine characteristics; "naïve experts" who wound up in trouble despite their brilliance; "evil plotters" who used their feminine wiles for nefarious ends; or "assistants" (often daughters) who helped male scientists.[54]

One notable, if circumscribed, exception came in 1997 with the portrayal of Ellie Arroway, the astronomer protagonist of *Contact*.[55] As Flicker observes, the movie depicts her character as highly competent but also shows her as surrounded by male colleagues who steal the limelight from her, male politicians who challenge her conclusions, and male funders on whom she is dependent to conduct her work.[56] Similarly, Jocelyn Steinke calls Arroway "a prototype of what a woman scientist role model should be ... an intelligent, capable, dedicated, committed, persistent, passionate, and successful scientist," while pointing out that the negative experiences the character faces in the movie—including workplace sexism—show "the frustration and disappointment that cause so many women to leave science."[57]

Like the gender gap in real-world science, the gender gap in Hollywood science has narrowed since *Contact* came out in 1997 but is still wide.[58] Analyzing scientists in 42 films made between 2002 and 2014, Steinke and Paolo Maria Paniagua Tavarez found that two-thirds of them were men whereas one-third were women.[59] The latter were especially likely to be biologists—not surprisingly, perhaps, given that the real-world gender gap is smaller in this discipline than in fields such as physics and engineering.[60] As for characterization, the films overwhelmingly tended to portray women scientists as professionals, rather than nerds, antisocial loners, or evil maniacs.[61] At the same time, virtually all of these characters were physically attractive, and almost a third of them were presented in a sexualized way.

Hollywood films from the past few years reflect the increasing prominence of women scientists as characters but also the enduring gender gap in movie science. For example, the only prominent scientist in *Jurassic World*—namely, geneticist Henry Wu—is a man. Likewise, a male botanist is the lead character in *The Martian*. The latter movie does feature several women scientists, including the mission commander and another astronaut on the mission, but the top three NASA administrators in the film are all men, as are the director of the Jet Propulsion Laboratory, the other three astronauts on the mission, and the astrodynamicist who calculates the flight trajectory that saves Watney. For now, women scientists such as the adventurous astronomer from *Pirates of the Caribbean: Dead Men Tell No Tales* (2017) and the evil chemist from *Wonder Woman* remain outnumbered by male counterparts such as the helpful seismologist from *San Andreas* (2015), the evil geneticist from *Logan* (2017), and even the heroic "magizoologist" from *Fantastic Beasts and Where to Find Them* (2016). The 2017 blockbuster *Jumanji: Welcome to the Jungle* offers a unique spin on this pattern when its four teenaged protagonists find themselves trapped in the bodies of video game avatars. The two young male leads become an adult male archeologist and an adult male zoologist, whereas one of the two young female leads receives the form of an adult *male* cartographer/paleontologist (the other young woman becomes a midriff-baring martial artist with no apparent scientific skills).

In addition to stereotyping scientists as men, Hollywood has tended to fill its white lab coats with white people. As Adilifu Nama writes, "American science fiction (SF) cinema has had a history of providing striking portrayals of the future, alternative worlds, sleek rocket ships, cyborgs, deadly ray guns, time machines, and wormholes through space, but, until quite recently, no black people."[62] Indeed, Weingart's study of 20th-century films found that fully 96% of the scientists in them were white.[63] During the early and middle parts of this era, films ranging from the disaster epic *When Worlds Collide* to the serious-minded and critically lauded *2001: A Space Odyssey* omitted any depiction of Black scientists.[64] Nor did Hollywood films typically include Latino or Asian scientists, with rare exceptions such as the Chinese-German mad scientist from the 1962 James Bond movie *Dr. No*—portrayed by a white actor in "yellow-face"—often falling into stereotypes of the sinister "other."[65] In effect, the movie industry replicated this era's real-life disparities in opportunities for people of color to pursue scientific careers.[66]

As attitudes among the public shifted and real-world science became more diverse, Hollywood began including more characters of color in its science-themed movies. However, these portrayals were typically limited to supporting roles. For example, Winston Zeddemore from *Ghostbusters* (1984) may be likable and heroic, but he is also the only Black member of the title team, the last one to join, and the only one who does *not* possess a scientific background. Likewise, the Black cyberneticist from *Terminator 2: Judgment Day* (1991) is a secondary character who ultimately dies as a "sacrificial lamb" to advance the story of the white leads.

The 2016 *Ghostbusters* reboot replaces the all-male original team with an all-woman team, but likewise portrays the one Black Ghostbuster as the sole non-scientist. Other recent films, however, have included greater racial and ethnic diversity in their portrayals of scientists. For example, the protagonists in the animated Disney film *Big Hero 6* (2014) include a young roboticist of Japanese and European descent, a Black laser expert, a Latina chemist, and an electromagnetic expert of Korean descent. Similarly, the lead scientist in *Jurassic World* is the Asian American Henry Wu. Meanwhile, the case of *The Martian* highlights both the trend toward greater inclusivity and lingering issues in representation. On the one hand, the movie features Black, Latino, and Asian American scientists in key roles (along with a white lead character and many other white scientists). On the other hand, some observers—including the Media Action Network for Asian Americans—criticized the filmmakers for not casting an actor of South Asian descent for the character of Vincent Kapoor (NASA's Director of Mars Missions, who is portrayed by a Black actor in the film) or an actor of Korean descent for the character of Mindy Park (a satellite expert, portrayed by a white actor in the film).[67]

While the film industry has traditionally excluded both women and people of color from its depictions of scientists, its record at portraying women of color as scientists is particularly limited. Given that women of color have long faced a "double bind" in real-world science from the intersecting forces of sexism and racism, this is an unfortunate case of art imitating reality.[68] However, two recent films go against Hollywood's pattern of ignoring women scientists of color.

One is *Hidden Figures*, which portrays three real-life Black women who worked for NASA during the 1960s: Katherine Johnson, Dorothy Vaughan, and Mary Jackson. The other is the 2018 box office smash *Black Panther*, in which the title hero's sister Shuri is the chief inventor for Wakanda, a fictional African nation that possesses technology superior to that of any other country in the world.

Beyond gender and race/ethnicity, Hollywood science has fallen short of inclusivity in other ways, as well. As a case in point, the dearth of LGBTQ scientists in popular movies reflects—and could reinforce—a historical pattern of invisibility within real-world science.[69] From the 1930s through the 1960s, the film industry's self-imposed censorship rules (known as the Production Code) forbade any overt depictions of LGBTQ people except as socially deviant villains. Some film scholars argue that *Bride of Frankenstein* director James Whale managed to sneak a gay scientist—albeit a mad one—past his era's censors in the form of the campy Dr. Pretorious.[70] However, it would be another four decades until the even campier Dr. Frank N. Furter could openly boast, "I've been making a man with blond hair and a tan, and he's good for relieving my tension." As recently as 2016, only 1% of the characters in the top hundred Hollywood films were LGBTQ characters. In keeping with this, the handful of science-themed movies featuring LGBTQ characters, such as the gay couple from 2017's *Alien: Covenant*, have tended to give their story arcs and romantic relationships relatively little focus. One exception is the biographical film *The Imitation Game* (2011), which portrays Alan Turing's foundational work in computer science along with the appalling treatment—including chemical castration—he suffered from the British government under its anti-LGBTQ policies. The movie was a financial success and received eight Academy Award nominations, including one for Best Picture.

Hollywood has a similarly bleak record when it comes to portraying disabled scientists. For the most part, the industry has restricted the roles of characters with physical disabilities to mad scientists such as the aforementioned Dr. No (who has prosthetic hands) and Dr. Strangelove (who uses a wheelchair). Another example of the same trope is Dr. Loveless from *Wild Wild West* (1999), an evil scientist who uses a wheelchair. Furthermore, the mad scientist archetype itself serves to reinforce stereotypes of people with mental illnesses as dangerous.[71] Even *A Beautiful Mind* (2001), which won an Academy Award for Best Picture, perpetuates myths and stereotypes about mental health in its sympathetic but dramatized portrayal of mathematician John Nash's schizophrenia.[72] To be sure, some recent films have featured disabled scientists in supporting roles as sympathetic antagonists or as assistants to the protagonists. For example, Dr. Connors from *The Amazing Spider-Man* (2012) is a well-intentioned scientist with a limb difference who inadvertently transforms himself into a lizard-person, while Dr. Gottlieb from *Pacific Rim* is a cane-using mathematician who helps to develop plans for battling giant monsters known as Kaiju. Despite these cases, disabled people (who make up almost 20% of the US population) remain underrepresented as movie scientists and as movie characters in general—just as they are underrepresented in actual scientific occupations.[73]

In part, Hollywood science's history of excluding women, people of color, LGBTQ people, and disabled people may reflect the barriers that members of these groups have faced in real-world science. Yet the demographics of Hollywood science may also reflect the demographics of the film industry itself. In a study of 900 recent films, Stacy Smith and her colleagues found that only 6% had Black directors, only 5% had women directors, and only 3% had Asian or Asian American directors (the study didn't look at how many directors were Hispanic, were LGBTQ, or had disabilities).[74] As for the double bind, Black women directed a grand total of 3 of the 900 films, while Asian or Asian American women directed two. The cases of *Wonder Woman* and *Black Panther* suggest that greater diversity among filmmakers could translate into greater diversity among scientist characters. A woman director, Patty Jenkins, helmed the former, which features a rare example of a woman as an evil scientist. A Black filmmaker, Ryan Coogler, directed the latter, which includes a Black woman in a prominent role dealing with science and technology.

This all matters because the visibility or invisibility of women scientists, scientists of color, LGBTQ scientists, and disabled scientists in fictional films may carry implications for how audience members perceive real-world science. If watching science-themed movies goes hand in hand with beliefs about what scientists are like, then watching such movies could also sway beliefs about who can be a scientist in the first place. Before children even reach adolescence, they begin developing mental models of how society works.[75] Most young people, and adults for that matter, have little firsthand interaction with scientists; as a result, media portrayals such as those found in Hollywood movies can loom large in their mental images of science.[76] When Hollywood portrays a lopsided majority of its scientist characters as white, able-bodied, heterosexual men—as the film industry has done throughout its history—then audience members may conclude that real-world science holds little place for those who don't fit this description. If, on the other hand, movies feature greater diversity in their depictions of scientists, then moviegoers may come away with a more inclusive vision of actual science.

This vision may be rosier than the present-day reality, but it could also help inspire change. Recent studies point to the importance of film portrayals in shaping public perceptions of who scientists are. For example, Jocelyn Steinke and her colleagues found that the middle school students they asked to draw a scientist frequently cited movies as inspirations for their pictures.[77] The STEM professionals whom Moria O'Keefe interviewed also described how seeing—or not seeing—film portrayals of scientists who looked like them influenced their impressions of the field. "Any images of Native Americans would have been helpful," one Native American engineer told her. "Images of Native people in contemporary life doing scientific work would have been amazing."[78] Likewise, a Black engineer mentioned her disappointment in how the 2008 film *City of Ember* portrayed all the "brightest minds in the world" as older white scientists.[79]

Research by Molly Simis and her colleagues suggests that Hollywood can present a more diverse image of science without sacrificing suspension of disbelief or willingness to buy tickets on the part of moviegoers. Box office data show that greater representation of women as movie characters in general does nothing to diminish overall box office returns and may enhance success in the United States.[80] Nor does featuring women scientists undermine audience acceptance of scientific plausibility. Far from it: when Simis and her team analyzed online discussions about a 2011 film featuring a woman in the key scientific role (*Thor*) and another 2011 film featuring a man in a similar role (*Rise of the Planet of the Apes*), they found that viewers saw the woman scientist as the more believable of the two.[81] The box office takes of *Hidden Figures* and *Black Panther* suggest that contemporary audiences are open to scientists who are Black women, as well. The same may be true for LGBTQ scientists and scientists with disabilities, particularly if Hollywood films were to include more of them. Our own 2018 poll found that most Americans favor greater inclusivity in science on all these dimensions, and diverse representations of the profession in movies could reinforce such support.[82]

From the Big Screen to the Small Screen

Though some in the scientific community see Hollywood as a font of inaccurate and negative portrayals, our look at science-themed films suggests a more complicated picture. By presenting perceptually realistic depictions of science and scientists, filmmakers can fuel misperceptions and reinforce stereotypes or boost and broaden the profession's image. On the one hand, movies such as *Jurassic World* present science that looks convincing on the screen but is controversial at best within the scientific community. On the other hand, the case of *The Martian* shows that filmmakers and science consultants can work together to produce movies that not only *seem* plausible but *are* realistic. Furthermore, such movies may help foster public support for endeavors such as sending astronauts to Mars. As for the classic mad scientist and nerd, they've increasingly given way to the scientist-as-hero. The stereotype of the scientist as a white, heterosexual, able-bodied man certainly persists on the silver screen, but a few recent movies offer models for greater diversity in Hollywood science.

Of course, film is not the only medium that presents images of science and scientists to the public—or even the most important purveyor of such images in the 21st century. As far back as fifty years ago, television supplanted movies as the dominant form of popular entertainment. Thus, it makes sense to ask whether television portrayals of science and scientists have followed the same trends as the ones in movies. It also behooves us to consider whether entertainment television has shaped public perceptions of these subjects—and, if so, how.

Notes

1 Information about box office returns and award nominations come from the Internet Movie Database (www.imdb.com).
2 Mooney, Chris, and Sheril Kirshenbaum, *Unscientific America: How scientific illiteracy threatens our future*, Basic Books, 2009, 82.
3 Jacobs, Tom, "The pervasive stereotype of the male scientist," *Pacific Standard*, Mar. 20, 2018, https://psmag.com/social-justice/pervasive-stereotype-of-the-male-scientist; Neill, Ushma S., "Hollywood's portrayals of science and scientists are ridiculous," *Scientific American*, Jan. 15, 2019, https://blogs.scientificamerican.com/observations/hollywoods-portrayals-of-science-and-scientists-are-ridiculous/.
4 Perkowitz, Sidney, *Hollywood science: Movies, science, and the end of the world*, Columbia University Press, 2007, 16.
5 Perkowitz, *Hollywood science*, 204.
6 Darren Naish, Nov. 25, 2014, https://twitter.com/TetZoo/status/537364767170899968.
7 Morgan Jackson, Nov. 25, 2014, https://twitter.com/BioInFocus/status/537300241 939439616.
8 Neil deGrasse Tyson, Oct. 2, 2015, https://twitter.com/neiltyson/status/6500898940 10687490.
9 Kirby, David A., *Lab coats in Hollywood: Science, scientists, and cinema*, MIT Press, 2011, 9; see also Grazier, Kevin R., and Stephen Cass., *Hollyweird science: From quantum quirks to the multiverse*, Springer, 2015.
10 Kirby, *Lab coats*, 33.
11 Kirby, *Lab coats*, 76.
12 Kirby, *Lab coats*, 69.
13 Kirby, *Lab coats*, 90.
14 McCarthy, Erin, "How NASA and Ridley Scott collaborated to make *The Martian*," *Mental Floss*, Oct. 2, 2015, http://mentalfloss.com/article/69351/how-nasa-and-ridley-scott-collaborated-make-martian.
15 Kutner, Max, "The scientist behind *Jurassic World*, Jack Horner, breaks down the movie's thrilling trailer," *Smithsonian Magazine*, Dec. 2, 2014, www.smithsonianmag.com/science-nature/scientist-behind-jurassic-world-breaks-down-trailer-180953505/.
16 Geggel, Laura, "Awesome dinos, iffy science inhabit *Jurassic World*," *Scientific American*, June 18, 2015, www.scientificamerican.com/article/awesome-dinos-iffy-science-inhabit-jurassic-world/.
17 Brown, Simon Leo, "*Jurassic World* dinosaur expert Jack Horner details where movies got the science wrong," ABC Radio Melbourne, Mar. 21, 2016, www.abc.net.au/news/2016-03-21/jurassic-world-dinosaur-expert-jack-horner-movies-vs-science/7263998; Kutner, "The scientist behind *Jurassic World*."
18 Kirby, *Lab coats*, 134, 150.
19 McCarthy, "How NASA."
20 Casey, Michael, "Paleontologists give *Jurassic World* science thumbs down," CBS News, June 15, 2015, www.cbsnews.com/news/paleontologists-give-jurassic-world-thumbs-down/.
21 Barnett, Michael, Heather Wagner, Anne Gatling, Janice Anderson, Meredith Houle, and Alan Kafka, "The impact of science fiction film on student understanding of science," *Journal of Science Education and Technology* 15, no. 2 (2006): 179–191.
22 Leiserowitz, Anthony A., "Before and after *The Day After Tomorrow*: A U.S. study of climate change risk perception," *Environment: Science and Policy for Sustainable Development* 46, no. 9 (2004): 22–39.

23 "Journey to Mars overview," NASA, June 30, 2018, https://www.nasa.gov/content/journey-to-mars-overview.

24 "Journey to Mars," NASA.

25 "Mars & beyond: The road to making humanity interplanetary," SpaceX, October 10, 2020, www.spacex.com/mars.

26 Bryce, Emma, "Neil deGrasse Tyson on alien life, Nasa's future and why he doubts humans will ever walk on Mars," *Wired*, Apr. 26, 2017, www.wired.co.uk/article/neil-degrasse-tyson-welcome-to-the-universe.

27 Aschwanden, Christie, "All we really need to get to Mars is a boatload of cash," *Five Thirty Eight*, Feb. 27, 2014, https://fivethirtyeight.com/features/all-we-really-need-to-get-to-mars-is-a-boatload-of-cash/; Kramer, Miriam, "Manned mission to Mars by 2030s is really possible, experts say," CBS News, Jan. 14, 2014, www.cbsnews.com/news/manned-mission-to-mars-by-2030s-is-really-possible-experts-say/.

28 Horner, John R., and James Gorman, *How to build a dinosaur: Extinction doesn't have to be forever*, Penguin, 2009.

29 Pilcher, Helen, *Bring back the king: The new science of de-extinction*, Bloomsbury Publishing, 2016.

30 Shapiro, Beth, *How to clone a mammoth: the science of de-extinction*, Princeton University Press, 2020.

31 Shultz, David, "Should we bring extinct species back from the dead?" *Science*, Sept. 26, 2016, www.sciencemag.org/news/2016/09/should-we-bring-extinct-species-back-dead.

32 "Woolly mammoth revival," Revive & Restore, Oct. 10, 2020, http://reviverestore.org/projects/woolly-mammoth/.

33 See the Appendix for details.

34 Weingart, Peter, Claudia Muhl, and Petra Pansegrau, "Of power maniacs and unethical geniuses: Science and scientists in fiction film," *Public Understanding of Science* 12, no. 3 (2003): 279–287.

35 Miller, Jon D., "Public understanding of, and attitudes toward, scientific research: What we know and what we need to know," *Public Understanding of Science* 13, no. 3 (2004): 273–294; Nisbet, Matthew C., Dietram A. Scheufele, James Shanahan, Patricia Moy, Dominique Brossard, and Bruce V. Lewenstein, "Knowledge, reservations, or promise? A media effects model for public perceptions of science and technology," *Communication Research* 29, no. 5 (2002): 584–608.

36 Skal, David J., *Screams of reason: Mad science and modern culture*, W W Norton & Company, 1998, 25–26.

37 Skal, *Screams of reason*, 315–316.

38 Haynes, Roslynn D., "Whatever happened to the 'mad, bad' scientist? Overturning the stereotype," *Public Understanding of Science* 25, no. 1 (2016): 31–44.

39 Haynes, "Whatever happened," 35.

40 Haynes, "Whatever happened," 41–42.

41 Grazier and Cass, *Hollyweird science*; Kirby, David A., "The changing popular images of science," in *The Oxford handbook of the science of science communication*, ed. Kathleen Hall Jamieson, Dan Kahan, and Dietram A. Scheufele, Oxford University Press, 2017: 291–300; Nisbet, Matthew C., and Anthony Dudo, "Entertainment media portrayals and their effects on the public understanding of science," in *Hollywood chemistry: When science met entertainment*, ed. Donna J. Nelson, Kevin R. Grazier, Jaime Paglia, and Sidney Perkowitz, American Chemical Society, 2013: 241–249.

42 Perkowitz, *Hollywood science*.

43 See Chapter 1.

44 Kirby, "The changing popular images"; Nisbet and Dudo, "Entertainment media portrayals."

45 Nisbet and Dudo, "Entertainment media portrayals."

46 Kirby, "The changing popular images."

47 Funk, Cary, and Brian Kennedy, "Public confidence in scientists has remained stable for decades," Pew Research Center, Mar. 22, 2019, www.pewresearch.org/fact-tank/2017/04/06/public-confidence-in-scientists-has-remained-stable-for-decades/.

48 Grazier and Cass, *Hollyweird science*; Haynes, "Whatever happened," Kirby, "The changing popular images"; Nisbet and Dudo, "Entertainment media portrayals"; Perkowitz, *Hollywood science*; Weingart et al., "Of power maniacs."

49 See Chapter 1.

50 Gross, Rachel E., "*The Martian* and the cult of science," *Slate*, Oct. 1, 2015, www.slate.com/articles/technology/future_tense/2015/10/ridley_scott_s_the_martian_film_science_worship_and_the_scientist_as_hero.html.

51 Latour, Bruno, *The pasteurization of France*, Harvard University Press, 1993.

52 Mooney and Kirshenbaum, *Unscientific America*, 82.

53 Weingart et al., "Of power maniacs."

54 Flicker, Eva, "Between brains and breasts—Women scientists in fiction film: On the marginalization and sexualization of scientific competence," *Public Understanding of Science* 12, no. 3 (2003): 307–318.

55 Steinke, Jocelyn, "Women scientist role models on screen: A case study of *Contact*," *Science Communication* 21, no. 2 (1999): 111–136.

56 Flicker, Eva, "Between brains and breasts."

57 Steinke, "Women scientist role models," 133.

58 For more on the real-world gender gap, see Chapter 1.

59 Steinke, Jocelyn, and Paola Maria Paniagua Tavarez, "Cultural representations of gender and STEM: Portrayals of female STEM characters in popular films 2002–2014," *International Journal of Gender, Science and Technology* 9, no. 3 (2018): 244–277.

60 National Science Board, "The state of U.S. science and engineering 2020," National Science Foundation/National Science Board, 2020. https://ncses.nsf.gov/pubs/nsb20201; Steinke and Taverez, "Cultural representations of gender and STEM."

61 Steinke and Taverez, "Cultural representations of gender and STEM." However, low-budget horror "B-movies" sometimes do depict women as mad scientists: see Jackson, J. Kasi, "Doomsday ecology and empathy for nature: Women scientists in 'B' horror movies," *Science Communication* 33, no. 4 (2011): 533–555.

62 Nama, Adilifu, *Black space: Imagining race in science fiction film*, University of Texas Press, 2010, 10.

63 Weingart et al., "Of power maniacs."

64 Nama, *Black space*.

65 Pua, Phoebe, and Mie Hiramoto, "Mediatization of East Asia in James Bond films," *Discourse, Context & Media* 23 (2018): 6–15.

66 For more on the real-world disparities, see Chapter 1.

67 Luhar, Monica, "*The Martian* faces accusations of 'whitewashing,'" NBC News, Oct. 12, 2015, https://www.nbcnews.com/news/asian-america/martian-faces-accusations-whitewashing-lead-characters-n442941.

68 Ong, Maria, Carol Wright, Lorelle Espinosa, and Gary Orfield, "Inside the double bind: A synthesis of empirical research on undergraduate and graduate women of color in science, technology, engineering, and mathematics," *Harvard Educational Review* 81, no. 2 (2011): 172–209.

69 For more on the latter pattern, see Chapter 1.

70 Skal, *Screams of reason*.

71 Harbour, Wendy, "*The Big Bang Theory*: Mad geniuses and the freak show of higher education," *Review of Disability Studies: An International Journal* 11, no. 2 (2015).

72 David, Anthony, "*A Beautiful Mind*," *British Medical Journal* 324, no. 7335 (2002): 491–492; Evans, Dominick, "Film critique: *Girl Interrupted* and *A Beautiful Mind*

depicting mental illness to society through film," *The Crip Crusader*, Feb. 6, 2015, www.dominickevans.com/2015/02/film-critique-girl-interrupted-and-a-beautiful-mind-depicting-mental-illness-to-society-through-film/.

73 Smith, Stacy L., Marc Choueiti, and Katherine Pieper, "Inequality in 1,300 popular films: Examining portrayals of gender, race/ethnicity, LGBTQ & disability from 2007 to 2019," Annenberg Inclusion Initiative, September 2020, http://assets.uscannenberg. org/docs/aii-inequality_1300_popular_films_09-08-2020.pdf; "Women, minorities, and persons with disabilities in science and engineering," National Science Foundation/National Science Board, 2017, www.nsf.gov/statistics/2017/nsf17310/static/data/tab9-8.pdf.

74 Smith, Stacy L., Marc Choueiti, and Katherine Pieper, "Inclusion in the director's chair: Gender, race, and age of directors across 1,200 top films from 2007 to 2018," Annenberg Inclusion Initiative, Jan. 2018, http://assets.uscannenberg.org/docs/inclusion-in-the-directors-chair-2007-2017.pdf.

75 Steinke, Jocelyn, "Cultural representations of gender and science: Portrayals of female scientists and engineers in popular films," *Science Communication* 27, no. 1 (2005): 27–63.

76 Simis, Molly J., Sara K. Yeo, Kathleen M. Rose, Dominique Brossard, Dietram A. Scheufele, Michael A. Xenos, and Barbara Kline Pope, "New media audiences' perceptions of male and female scientists in two sci-fi movies," *Bulletin of Science, Technology & Society* 35, no. 3–4 (2015): 93–103; Steinke and Taverez, "Cultural representations of gender and STEM."

77 Steinke, Jocelyn, Maria Knight Lapinski, Nikki Crocker, Aletta Zietsman-Thomas, Yaschica Williams, Stephanie Higdon Evergreen, and Sarvani Kuchibhotla, "Assessing media influences on middle school–aged children's perceptions of women in science using the Draw-A-Scientist Test (DAST)," *Science Communication* 29, no. 1 (2007): 35–64.

78 O'Keeffe, Moira, "Lieutenant Uhura and the drench hypothesis: Diversity and the representation of STEM careers," *International Journal of Gender, Science and Technology* 5, no. 1 (2013): 4–24.

79 O'Keeffe, "Lieutenant Uhura," 12.

80 Simis et al., "New media audiences' perceptions."

81 Simis et al., "New media audiences' perceptions."

82 See Chapter 1.

3

PRIME-TIME SCIENCE

Sheldon Cooper: I'm about to embark on one of the greatest challenges of my scientific career: teaching Penny physics.

—*The Big Bang Theory* (2009; Season 3, Episode 10: The Gorilla Experiment)

Penny: You do your little experiments, I do mine.

—*The Big Bang Theory* (2007; Season 1, Episode 8: The Grasshopper Experiment)

In the fall of 2007, the CBS television network launched a new sitcom named after one of the most important concepts in astrophysics: *The Big Bang Theory*. The show originally featured five main characters, all of whom remained on the program through its 12-season run. Four of these are classic examples of the geeky male scientist. Experimental physicist Leonard Hofstadter, the least nerdy of them, often makes snarky comments about the others' quirks. Meanwhile, theoretical physicist Sheldon Cooper embodies popular stereotypes of scientists as strange (he's germophobic and punctuates his conversations with the catchphrase "Bazinga!"), socially awkward (he's an insufferable know-it-all), and unfashionable (the "short-sleeved T-shirt over long-sleeved T-shirt" look is practically his uniform). The others, aerospace engineer Howard Wolowitz and particle astrophysicist Rajesh (Raj) Koothrappali, occupy a middle ground of geekiness between Leonard and Sheldon. Penny, who lives across the hall from Leonard and Sheldon, is both the only non-scientist and the only woman among

DOI: 10.4324/9781003190721-3

the original main cast of characters (as well as the only character whose last name is never revealed). She serves as a foil for the male scientists and a source of "ditzy blonde" humor.

From its first season onward, the show's producers strove to include accurate portrayals of science so that audience members would perceive its premise as realistic, suspend their disbelief, and focus on the characters' antics.[1] To this end, they hired particle physicist David Saltzberg as a science consultant. He provided them with advice on many aspects of the show, from the look of its characters' labs to the equations they solve on white boards to the jargon they use. Intentionally or not, the program also replicated the real-world gender disparity in science occupations by casting all of the main physicists as men. As Saltzburg himself told one interviewer in 2008, the ratio of women to men in his field "is pretty bad."[2]

The Big Bang Theory gradually developed into a television hit, becoming the top-rated prime-time sitcom during its fourth season.[3] That season also saw the promotion of two recurring characters to the main cast: Amy Farrah Fowler, a neurobiologist and love interest for Sheldon, and Bernadette Rostenkowski, a microbiologist and love interest for Howard. During its seventh season, the program earned the second-highest ratings of any 2013–2014 prime-time program. It went on to repeat this performance for the next three seasons.

The long-running success of *The Big Bang Theory* highlights the role of entertainment television in propagating popular images of scientists. This role extends all the way back to the infancy of commercial television in the early 1950s, which featured children's science fiction shows such as *Captain Video and His Video Rangers* and *Tom Corbett, Space Cadet*. As television became *the* dominant form of media in the late 1950s and 1960s, several enduring classics debuted on the airwaves, including *The Twilight Zone, Star Trek,* and British import *Doctor Who*. From the 1970s to the 1990s, scientists appeared in prime-time network programs ranging from space operas such as the original *Battlestar Galactica* to paranormal dramas such as *The X-Files*. Fictional scientists in the more fragmented media landscape of recent years have included not only the wacky sitcom characters of *The Big Bang Theory* but also the crime-solving forensic scientists of the *CSI* franchise, *NCIS,* and *Bones* and the sometimes-heroic, sometimes-mad scientists of science fiction programs such as *Lost, Stranger Things,* and *Marvel's Agents of S.H.I.E.L.D.*

These television scientists loom large in the public imagination. In October 2016, we asked a nationally representative sample of 1,000 Americans to name the first person who came to mind when they thought about scientists on TV shows or in movies.[4] Of the respondents, 17% mentioned characters from fictional television programs. Almost one in ten (9%) cited characters from *The Big Bang Theory* alone. Sheldon received by far and away the most mentions, followed by Leonard. Respondents also name-checked Howard and Raj, and even bit character Professor Proton, though not one specifically mentioned Amy or Bernadette.

No other program came close to *The Big Bang Theory*'s number of mentions, but respondents did name a wide variety of shows. Collectively, the fictional scientists from the *Star Trek* franchise came in second place (2%)—due in no small part to Spock, the U.S.S. *Enterprise*'s Vulcan science officer. The forensic scientists on *Bones* (most notably, forensic anthropologist Temperance Brennan; 1%), *NCIS* (particularly Abby Sciuto; 1%), and the *CSI* franchise (many characters, though respondents mentioned two by name: Gil Grissom and Nick Stokes; 1%) accounted for the next three spots in the rankings. Respondents also cited chemistry-teacher-turned-meth-cook Walter White from *Breaking Bad*, skeptical Federal Bureau of Investigation (FBI) agent Dana Scully from *The X-Files*, lovelorn paleontologist Ross Gellar from *Friends*, dorky Professor Frink from *The Simpsons*, and acid-dropping Walter Bishop from *Fringe*, among many others.

The prominence of fictional television scientists in the public's imagination raises the possibility that watching entertainment television will shape how viewers perceive science. For example, frequent exposure to positive portrayals could sway audience members to see scientists as being dedicated to the good of humanity, while repeated viewing of scientists being hurt or killed could lead audience members to perceive scientific work as dangerous. Such effects, in turn, may ultimately matter for whether members of the public support scientific endeavors, trust what scientists say about important issues, and engage in citizen science. These portrayals could also influence whether young people choose to pursue scientific training and occupations: viewers might be inspired by depictions of scientists as good but discouraged by images of the profession as hazardous. By the same logic, prime-time shows that reinforce or challenge common stereotypes of the profession could affect audience members' beliefs about what scientists look like. As a case in point, the gender balance among prime-time scientists could shape perceptions of real-life gender disparities in science—and, perhaps, beliefs about who can be a scientist. Beyond the impact of overall television viewing, particular genres such as science fiction television and even individual programs such as *The X-Files* and *The Big Bang Theory* could influence perceptions of what life as a scientist is like and whether scientists are relatable role models, socially awkward oddballs, or a bit of both.

To explore these potential effects, we'll use a framework that has driven half a century of research on entertainment television: cultivation theory.

Cultivation Theory and Science

In the 1960s, George Gerbner developed an approach for understanding how the increasingly popular medium of television might influence the public.[5] Cultivation theory, as proposed by Gerbner and his colleagues, posits that dominant messages within television programming shape viewers' perceptions of reality.[6] In essence, the theory argues that television has taken the place of folklore and fables as the repository of society's shared stories.

Some of Gerbner's earliest and most influential work focused on television violence. Following his 1968 appointment to the National Commission on the Causes and Prevention of Violence, he began a long-running study of television content. Gerbner and his team found that prime-time television programming was rife with portrayals of violence. Furthermore, they concluded that a steady diet of exposure to such portrayals could fuel perceptions of a hostile, dangerous society—a "mean world syndrome." In keeping with this, they found links between higher levels of television exposure and a host of outcomes, including greater distrust of people and heightened fears of becoming a victim of violence.

Cultivation theory has inspired considerable debate over the years.[7] For example, its critics have argued that it neglects important differences across television genres.[8] Even so, the theory provides a useful launching point for investigating how prime-time television depicts scientists. In fact, Gerbner and his colleagues conducted a series of studies doing just that.[9] Looking at the period from 1969 to 1983, they found that prime-time programs frequently included scientific themes but seldom included scientists as characters: leaving aside medical doctors, the average viewer saw only one or two scientists on television per week. The researchers also discovered an ambivalent portrayal of these scientists. On the one hand, prime time depicted them as five times more likely to be virtuous than villainous. On the other hand, prime time also portrayed scientists as particularly strange and unsociable. Furthermore, they suffered injury or death at a higher rate than any other occupational group.

Building on these findings, Gerbner and his team used data from a 1983 survey to argue that higher levels of television viewing fostered negative views of science, including skepticism about new technologies, willingness to place restrictions on science, and stereotypes of scientists. For example, heavy viewers were especially likely to see scientists as peculiar and scientific work as dangerous. "Foreboding images of [science as an] odd and perilous activity," Gerbner wrote, "seem to heighten fears, strengthen the desire for restraint, and inhibit the inclination for science as an occupation or an area for public participation."[10]

Subsequent research, however, has suggested that entertainment television portrayals of science and scientists can have a double-edged impact on public perceptions.[11] Looking at survey data from 1999, Matthew Nisbet and his colleagues found that television viewing still promoted *reservations* about science and technology, partly by presenting science as "weird and frightening" and partly by taking up time that viewers might otherwise have spent learning about it.[12] Yet television viewing simultaneously promoted belief in the *promise* of science by portraying science and scientists as "powerful forces for good."[13]

More recently, Anthony Dudo and his collaborators extended research on cultivation and perceptions of science to the first decade of the 21st century. As in previous decades, scientists seldom appeared on prime time during this period. When they did, they tended to be more good than otherwise (by a four to one ratio) but were particularly likely to be the victims of violence.[14] The researchers' analysis of 2006 survey data revealed that television viewing played no direct role in promoting negative views of science, though it played an indirect one through the same process of time displacement that Nisbet and his colleagues observed.

In short, previous studies suggest that mixed messages about science and scientists on entertainment television can produce a range of effects on viewers' perceptions—some negative, some positive. Here, we take a new look at such effects. In doing so, we keep in mind that prime-time television programs, like Hollywood films, have shifted increasingly toward portraying scientists as heroes instead of mad villains or eccentric pariahs.[15]

Prime-Time Scientists

To provide a more recent portrait of scientists on prime-time television, we draw on data from content analysis research conducted by Michael Morgan, James Shanahan, and Nancy Signorielli for the Cultural Indicators Project founded by Gerbner.[16] This project, which spanned five decades, captured how entertainment television programs portrayed different sorts of characters over time. Looking at a sample of episodes across major networks from each year, the researchers examined every character who played a key role in the program's action.

All told, they analyzed 11,556 characters who appeared on prime-time television between 1973 and 2015. Of these, only 1% were scientists. By comparison, six times as many characters were medical professionals and ten times as many were police officers or private investigators. Put simply, scientists were relatively rare on prime-time programs. In the 2010s, the proportion of scientists as major characters doubled from previous decades—all the way to 2%, still hardly an overwhelming presence.

In terms of their morality and ethics, fictional television scientists have tended to be good, rather than evil or somewhere in between. From 1973 to 2015, prime time portrayed 71% of all scientists as primarily good, only 6% as bad, and 23% as a mix of each. Indeed, scientists were more likely to be good than characters in general (60% of whom were good). Scientists on prime time have become more heroic over time, as well. From the 1970s through the 1990s, around half of all scientists were good, versus half who were bad or mixed. By the first decade of the 2000s, good scientists substantially outnumbered bad or mixed ones, 78% to 22%. From 2010 onward, scientists were almost universally good: 94%, versus 6% who were bad or mixed.

The answers to the "name a scientist" question in our 2016 survey reflect the prevalence of good scientists on prime-time television. Some of the scientists our respondents named are action heroes such as Cisco Ramon from *The Flash* or Samantha Carter and Rodney McKay from *Stargate SG-1*. Others are forensic scientists who solve crimes, such as Gil Grissom and Nick Stokes from *CSI*, Temperance Brennan from *Bones*, and Abby Sciuto from *NCIS*. Dana Scully, one of the alien-and-monster-chasing FBI agents from *The X-Files*, is a bit of both. Most of the other scientists mentioned by our respondents are ordinary but fundamentally benign people, including the scientists from *The Big Bang Theory* and Ross Gellar from *Friends*.

To be fair, respondents did name a few morally ambiguous or downright villainous prime-time scientists. Walter Bishop from *Fringe* fits the first category: throughout the series, he strives to redeem himself after conducting highly unethical experiments on children to give them psychic abilities. Walter White follows the opposite trajectory on *Breaking Bad* as he descends into a life of crime as a drug lord. Another character, Harrison Wells from *The Flash*, initially acts as a mentor to the show's hero but turns out to be a supervillain from the future. Still, sinister scientists are the exception, rather than the rule, on contemporary entertainment television.

In terms of their involvement in violence, prime-time scientists were no more likely to *commit* acts of violence (29% of them did so) than characters in general (31%). Meanwhile, prime time portrayed scientists as particularly likely to be the *victims* of violence. Across the five decades, 38% of scientists were hurt or killed, compared to 31% of all characters. Prime time was especially hazardous for fictional scientists prior to 2000; during this period, around half of all scientists were hurt or killed. The early 2000s offered a relative reprieve, with 20% hurt or killed, but the casualty rate ticked up in the 2010s to 32%.

The fictional television scientists mentioned by the respondents in our October 2016 survey included both those who are strangers to violence and those who live dangerously. In no small part, these characters' propensity to suffer harm reflects the genres in which they appear. On sitcoms, scientists can generally count on not being the victims of assault or murder. For example, Leonard from *The Big Bang Theory* doesn't have to worry about terrorists trying to blow up his laboratory, just as Ross from *Friends* is never chased down and eaten by a live Tyrannosaurus Rex.

By contrast, characters on forensic crime dramas such as *Bones* and *CSI* not only solve violent crimes but also risk becoming crime victims themselves. For example, a vengeful kidnapper buries Nick Stokes alive in one episode of *CSI* (his colleagues rescue him). Similarly, an episode of *Bones* begins with someone shooting Temperance Brennan in her lab (she survives). Scientists on other sorts of programs also experience violence, ranging from the slapstick sufferings of lab assistant Beeker on *The Muppets Show* to the machine gun wounds received by

Walter White in the finale of *Breaking Bad*. Beeker always recovers, but White is not so fortunate.

Cultivating Perceptions of Science

To examine links between television viewing and contemporary perceptions of science and scientists, we used data from a nationally representative survey of 900 US residents that we conducted in July 2016.[17] This survey asked respondents how many hours of television they watched on an average day, including viewing on a computer, tablet, or phone. It also included two questions about science and scientists. One asked respondents how much they agreed or disagreed that "scientists are dedicated people who work for the good of humanity." The other asked how much they agreed or disagreed that "scientific work is dangerous."

We found clear relationships between levels of television viewing and answers to each question (Figure 3.1). First, take respondents' views on whether scientists

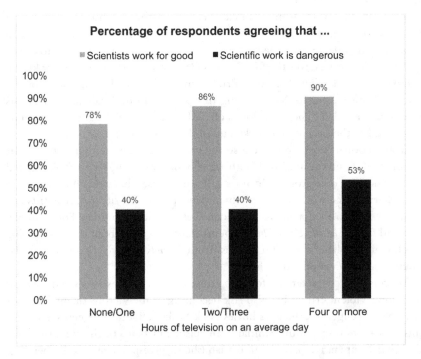

FIGURE 3.1 Perceptions of scientists and scientific work, by overall TV watching (Center for Political Communication Survey, 2016)

are dedicated people who work for good. A sizable majority (78%) of those who watched zero to one hour of television a day agreed with this statement. However, the respondents who watched at least four hours of television were even *likelier* to agree: fully 90% of them did so. This pattern fits with the increasing tendency of prime-time television to portray scientists as good.

Respondents who watched four or more hours of television a day also stood out in terms of how dangerous they saw science as being. Slightly more than half (53%) of them agreed that scientific work is dangerous, compared to only 40% of respondents who watched zero to one hour. This pattern, in turn, dovetails with prime time's ongoing portrayal of scientists as disproportionately likely to suffer injury or death.

Both relationships remained statistically significant even after we controlled for a host of other potentially relevant factors, including respondents' background characteristics and other forms of media use. Thus, our results reinforce the point that exposure to entertainment television depictions of science can carry implications for viewers' perceptions. In recent years, such portrayals have promoted an image of scientists as good, but also of science as perilous. The first pattern could lead viewers to trust scientists and even to consider pursuing scientific careers, while the second could discourage engagement with science. Yet these patterns may dovetail with one another, too: some prime-time scientists are heroes precisely because they face danger to help people.

Science Fiction TV and Perceptions of Science

The original version of Gerbner's cultivation theory focused on television as a whole, rather than particular genres—let alone particular programs. Subsequent research, however, has highlighted the potential for one specific genre to shape perceptions of science: science fiction, which emphasizes the dramatic possibilities of speculative science.[18] For example, Matthew Nisbet and Robert Goidel showed that people who regularly watched science fiction programs expressed particularly strong support for biomedical research on cloning.[19] Then again, Dominique Brossard and Anthony Dudo found no evidence that science fiction television viewing directly influenced broader perceptions of science.[20] Thus, one possibility is that science fiction shows influence some perceptions of science but not others.

The content analysis for the Cultural Indicators Project didn't compare science fiction programs to other prime-time programs. Nevertheless, it seems likely that recent science fiction shows, like recent entertainment television shows in general, have tended to portray scientists as predominantly good. Of the science fiction characters named by respondents in our 2016 online survey, most—such as Spock from *Star Trek*, Dana Scully from *The X-Files*, and Samantha Carter from the *Stargate* franchise—follow this pattern, though there are exceptions such as the evil Harrison Wells from *The Flash* and the conniving Dr. Smith from *Lost in Space*.

Given that science fiction programs tend to be action-oriented dramas, they may also be particularly likely to portray science as a dangerous endeavor. Whereas a life of science seldom carries any risk on *The Big Bang Theory*, scientists often face great peril working for Starfleet from *Star Trek*, Stargate Command from *Stargate*, the FBI's Fringe Division from *Fringe*, or S.T.A.R. Labs from *The Flash*. Just ask Spock, whose entire brain is stolen on one episode of *Star Trek*, or Walter Bishop from *Fringe*, who merely has *pieces* of his brain removed.

Marvel's Agents of S.H.I.E.L.D. illustrates how science fiction television shows present scientists as good while also presenting science as dangerous. This ABC program, which premiered in 2013 as a spinoff from a franchise of superhero movies (the Marvel Cinematic Universe), focuses on a team of agents who fight shadowy terrorist organizations, villains with superpowers, and warlike aliens. Two key members of the team are Fitz Leopold, a brilliant engineer, and Jemma Simmons, an equally brilliant biochemist. Working in tandem, they provide scientific and technological solutions to many of the team's problems. For example, in one episode Fitz creates a "quantum field disruptor" to trap an opponent who possesses the ability to teleport, then triumphantly proclaims, "Science, biatch!" in his Scottish accent.[21] In another episode, Simmons uses her scientific skills to stay alive on a deadly alien planet for almost 200 days. Yet working for S.H.I.E.L.D. carries grave risks—as Fitz learns firsthand when he suffers a brain injury from which it takes him months to recover, and as Simmons experiences during her long, traumatic ordeal on the alien world.

The BBC series *Doctor Who*, which has devoted fan bases on both sides of the Atlantic, shows the same patterns. During its half-century run, it has depicted literally hundreds of scientists.[22] Many of these characters are admirable and even heroic, including geoscience engineer Nasreen Chaudhry, astronomer Dee Blasco, and Kate Stewart of the intelligence organization UNIT (whose credo is "science leads").[23] Of course, they need to be, given that they often face menaces such as the Cybermen (aliens who seek to remake other species in their own image) and the Daleks (aliens bent on exterminating other forms of life). Sometimes the scientists of *Doctor Who* even die at the hands of the villains: for example, UNIT scientist Petronella Osgood is slain by an incarnation of the evil Time Lord known as the Master.

To test whether science fiction television viewing is linked to perceptions of science, we revisited the data from our July 2016 survey. In addition to measuring overall television viewing, this survey asked respondents how often they watched specific types of programming, including "science fiction television shows such as *Doctor Who* or *Marvel's Agents of S.H.I.E.L.D.*" We found mixed results (Figure 3.2). On the one hand, respondents who sometimes or regularly

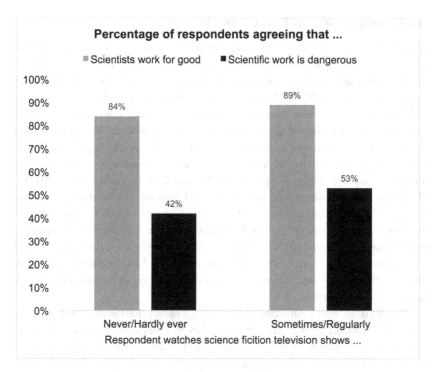

Percentage of respondents agreeing that ...

■ Scientists work for good   ■ Scientific work is dangerous

FIGURE 3.2 Perceptions of scientists and scientific work, by science fiction TV watching (Center for Political Communication Survey, 2016)

watched science fiction programs and those who never or hardly ever watched such programs differed relatively little on whether scientists work for good: 89% versus 84%. The absence of a clear link here fits with the finding that almost *all* prime-time portrayals of scientists have been positive throughout the 2010s. On the other hand, science fiction viewers were substantially more likely than non-viewers to see scientific work as dangerous. A majority (53%) of those who regularly or sometimes watched science fiction agreed with this idea, compared to a minority (42%) of never-or-hardly-ever viewers. Such a pattern makes sense if science fiction programs are especially prone to portray scientists coming to harm.

Similar results emerged even after we controlled for a range of other factors, including overall television viewing and other forms of media use. In terms of genre-specific viewing, then, we found that science fiction television exposure helps explain perceptions of science as dangerous but not of scientists as good. More broadly, our results suggest that genre-specific viewing may matter most when the genre in question presents science in a distinctive way.

The Demographics of Prime-Time Science

Along with shaping perceptions of *what* science and scientists are like, entertainment television may also shape perceptions of *who* scientists are. Such an impact could be especially important given that women, people of color, LGBTQ people, and disabled people are all underrepresented in real-world science.[24]

A look at *The Big Bang Theory* reveals how prime-time television can mirror such disparities. Of the six main scientists on the show from the fourth season onward, four are men and only two are women. The percentages here closely reflect the gender gap in the actual science and engineering workforce: 67% versus 33% on the show, compared to 71% and 29% in reality (as of 2017).[25] The program also mirrors another real-world pattern: the relative prevalence of women in the life sciences and their relative absence from the physical sciences and engineering. In 2017, 48% of the people in life sciences occupations were women, compared to 29% for the physical sciences, 27% for computer and mathematical sciences, and 16% for engineering. On *The Big Bang Theory*, both Amy and Bernadette are biologists whereas the four men in the main cast are all physicists or engineers.

In terms of race and ethnicity, *The Big Bang Theory* does even worse than the real world. Thanks to Raj, the show roughly mirrors the proportion of Asian scientists and engineers in the US (one in five). Then again, the other five scientists on the show are all white characters: 83%, versus 65% in the real world. None of the main scientists on *The Big Bang Theory* is Black or Hispanic (as of 2017, 6% of real-life scientists and engineers were Black and 8% were Hispanic). Nor are any recurring minor scientist characters, though astrophysicist Neil deGrasse Tyson (who is Black) plays himself in an episode where he argues with Sheldon about the demotion of Pluto from planet to "dwarf planet."

The program doesn't feature any LGBTQ scientist protagonists, either. When *The Big Bang Theory* addresses LGBTQ identities through its main characters, it tends to do so through "gay panic" jokes about other characters incorrectly inferring a romantic relationship between Howard and Raj. In one episode, Raj's parents think he's "coming out" as gay to them; he tells them, "No, I'm not gay. If anything, I'm a metrosexual ... I like woman, as well as their skin care products."[26] Here, then, the program reflects some of the stereotyping and invisibility that LBGTQ people face in real-world science.[27]

In terms of disability and *The Big Bang Theory*, many viewers—and some disability rights activists—have assumed that Sheldon has an autism spectrum disorder or obsessive-compulsive personality disorder given behaviors such as his ritualistic knocking and aversion to germs.[28] The program never explicitly confirms this in any episode; nor have its creators endorsed the premise in interviews.[29] However, Jim Parsons (the actor who plays Sheldon) has lent support to the idea that Sheldon has an autism spectrum disorder. For her part, Mayim

Bialik (who portrays Amy and holds a real-life neuroscience Ph.D.) has suggested a diagnosis of obsessive-compulsive personality disorder. One episode from the program's sixth season shows her character giving Sheldon "a series of exercises to help with your compulsive need for closure."[30] Leaving aside the ambiguous case of Sheldon, the show's most prominent portrayal of a scientist with a disability may have come in the form of several cameos by theoretical physicist Stephen Hawking, who used a wheelchair in real life due to amyotrophic lateral sclerosis (ALS).

For a more comprehensive portrait of prime-time scientists in terms of sex and race, we revisited the data from the Cultural Indicators Project. As it turns out, *The Big Bang Theory* is no fluke when it comes to the gender gap on prime-time science. Across the five decades of the study, 68% of the fictional scientists on television were men and 32% were women. This imbalance exceeded the overall gender imbalance on prime time, where 62% of all characters were men. The gender ratio among fictional scientists remained roughly constant from the 1970s to the early 2000s: in each decade, men outnumbered women by at least a two-to-one margin. The gap narrowed in the 2010s, when women scientists such as Abby Sciuto (*NCIS*) and Temperance Brennan (*Bones*) occupied key roles on popular programs. Even in this decade, however, 58% of prime-time scientists were men and only 42% were women.

The results of our October 2016 survey mirror the gender gap in prime-time science. Most of the respondents who named a specific fictional television scientist chose a male one. Some of these were characters from older programs, such as Spock from *Star Trek*, the Professor from *Gilligan's Island*, and Dr. Smith from *Lost in Space*. Others came from recent programs, including the men from *The Big Bang Theory*, Walter White from *Breaking Bad*, and the male forensic scientists from *CSI*. The relatively few respondents who named a specific woman tended to choose either Abby Sciuto from *NCIS* or Temperance Brennan from *Bones*. Only three other women television scientists received any mention at all: Beverly Crusher from *Star Trek: The Next Generation*, Dana Scully from *The X-Files*, and Samantha Carter from *Stargate SG-1*.

Just as prime-time scientists have tended to be male, they've also tended to be white. In part, this reflects the broader predominance of white characters on prime time. From 1973 to 2015, 84% of all characters were white, compared to 12% who were Black and 4% who were of another race (the Cultural Indicators Project didn't code whether characters were Hispanic). If anything, however, the disparity was especially marked among scientists. Fully 89% of prime-time scientists were white, compared to 7% who were Black and 4% who were of another race. Furthermore, prime-time science hasn't grown more racially diverse over time. Far from it: the percentage of prime-time scientists who were white went from 80% in the 1970s and 77% in the 1980s to 89% in the 1990s, 96% in the 2000s, and 91% in the 2010s.

Again, the results of our 2016 survey reflect the demographics of prime-time science. Almost all of the fictional television scientists whom respondents mentioned were white: Sheldon, Leonard, and Howard from *The Big Bang Theory*, Abby Sciuto from *NCIS*; Temperance Brennan from *Bones*; Gil Grissom and Nick Stokes from *CSI*; Walter White from *Breaking Bad*; Dana Scully from *The X-Files*; the Professor from *Gilligan's Island*, Ross Gellar from *Friends*, and so on. Respondents mentioned precisely one Asian American character (Raj of *The Big Bang Theory*, who is an immigrant from India), one Hispanic character (Cisco Ramon from *The Flash*, who is Puerto Rican), and zero Black characters.

Though the Cultural Indicators Project didn't record characters' sexual identity or disability status, we know from other studies that prime-time television rarely features LGBTQ characters or disabled characters—scientists or otherwise. A 2017 study by GLAAD found that LGBTQ characters made up only 5% of all prime-time characters, and a 2017 study by the Ruderman Family Foundation found that only 2% of television characters had a disability.[31] Characters with multiple marginalized identities, such as LGBTQ characters of color and LGBTQ characters with disabilities, are even scarcer on prime-time programs.[32] Thus, it is no surprise that few of our survey respondents named disabled scientists (unless one counts Sheldon from *The Big Bang Theory*) or LGBTQ scientists. One exception was the lone respondent who mentioned Dr. Swann from *Smallville*, an astronomer who uses a wheelchair (portrayed by Christopher Reeve, who used one in real life).

Perceptions of Who Scientists Are

Cultivation research suggests that the demographics of entertainment television matter. Specifically, watching television can mold viewers' perceptions of social groups in the real world. For example, cultivation-based research has found links between television viewing habits and perceptions of gender roles.[33] Likewise, television watching habits help explain perceptions of marginalized racial groups, LGBTQ people, and disabled people.[34]

We drew on data from our November 2016 national survey to test whether this pattern extends to perceptions of real-world gender ratios in STEM professions.[35] The study asked respondents how many hours a day of entertainment television they watched on average. It also asked them to estimate the percentages of men and women working in six fields: biology, chemistry, physics, astronomy, engineering, and computer science. On average, respondents moderately underestimated the percentage of biologists who were women: 44%, versus 52% in the real world.[36] Meanwhile, they slightly overestimated the percentage of chemists who were women: 39%, compared to the actual figure of 35%. They erred on the high side for astronomy, as well: 36% versus 29%. For the other fields, respondents substantially overestimated the representation of women: 41% versus 26% for computer science, 34% versus 17% for physics, and 35% versus 15% for engineering.

In terms of cultivation effects, watching more entertainment television went hand in hand with higher estimates for the percentage of women working in STEM—even after controlling for the viewer's own gender and other background factors (Figure 3.3). Compared to respondents who watched no entertainment television, those who watched four or more hours a day offered estimates that averaged two percentage points higher for physics; four points higher for biology, astronomy, and engineering; five points higher for chemistry; and six points higher for computer science. Though these differences were not especially large, they were statistically significant for five of the six fields (physics being the lone exception).

Such findings raise a question: how could entertainment television viewing contribute to *overestimating* the percentage of women in real-world science

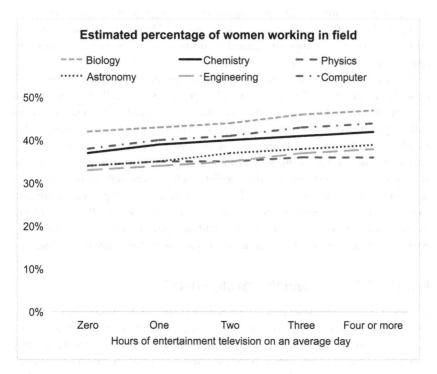

FIGURE 3.3 Perceptions of gender ratios in scientific fields, by entertainment TV watching (Cooperative Congressional Election Survey, 2016)

and engineering occupations, when men still outnumber women as prime-time scientists? One potential explanation follows from the finding that the gender gap in science has narrowed more quickly on prime time than it has in the real world. On prime time, scientists went from a two-to-one gap in the 1990s to better than a three-to-two gap in the 2010s. Meanwhile, real-life science and engineering went from around a three-to-one gap in the 1990s to a bit worse than a two-to-one gap in the 2010s. Thus, entertainment television may leave viewers with the impression that the real-world gender balance in science is more equitable than it truly is.

The November 2016 survey didn't include questions asking respondents to estimate the proportions of scientists and engineers who were people of color, who were LGBTQ, or who had disabilities. As a result, we can't test for links between television viewing habits and perceptions of these proportions. Given the dearth of nonwhite scientists on prime time, however, it seems plausible that heavy levels of entertainment television viewing would go hand in hand with lower estimates for the percentages of nonwhite scientists and engineers in the real workforce. Similarly, the invisibility of LGBTQ scientists and disabled scientists on prime time could lead frequent television viewers to assume that few, if any, real-world scientists share these identities. Unless the demographics of prime-time science change dramatically, future researchers will have plenty of opportunities to test our speculations here.

We should also point out that counting the proportions of television scientists who belong to different demographic groups doesn't tell the full story of representation for these groups. If prime time also *characterizes* scientists differently depending on their group identities, then even depictions that are equitable based on numbers may still reinforce stereotypes. For example, a tendency to portray women as life scientists and men as physical scientists—as on *The Big Bang Theory*—could strengthen the associations that viewers draw between gender and specific scientific fields.[37] Likewise, a tendency to portray Asian and Asian American scientists as unsuccessful at romance—as *The Big Bang Theory* did in its early years with Raj—could reinforce intersecting stereotypes about Asian and Asian American men as nerdy *and* as less traditionally masculine.[38]

Parasocial Contact and the "Scully Effect"

If both entertainment television as a whole and specific television genres influence how viewers perceive science, then perhaps individual television programs can do the same. In particular, a growing body of research demonstrates that "parasocial" interactions between viewers and television characters can sometimes produce effects similar to those of real-world social interactions.[39] For example, one study found that watching the television sitcom *Will & Grace*, which features two gay lead characters, reduced anti-LGBTQ prejudice among viewers.[40] Another study found that forming a parasocial relationship with the title character of the television series *Monk*, who has obsessive-compulsive disorder, went hand in hand with holding positive views of people with the same disorder.[41] So, how might parasocial contact with television characters influence what audience members think about scientists?

In the case of *The X-Files*, the character of FBI agent Dana Scully—as portrayed by Gillian Anderson—may have influenced young people, particularly young women, to hold more positive views of science and to see themselves as potential scientists. "We got a lot of letters all the time," recalled Anderson. "And I was told quite frequently by girls who were going into the medical world or the science world or the FBI world or other worlds [where] I reigned, that they were pursuing those pursuits because of the character of Scully."[42]

The program's science consultant, virologist Anne Simon, saw the character as providing many of her own students with a role model. "I asked my Intro Bio class ... how many of them were influenced by the character of Scully on *The X-Files* to go into to science," she said, "and half the hands in the room went up ... I think yes, there was a Scully effect."[43] Simon also recounted the story of a high school student who told her, "I wanted to become a scientist, but I'm not a nerd. I'm not a geek. I'm not ugly. I'm not mad. And then I started watching *The X-Files* and I saw Scully—and she's beautiful, and she's smart, and she's believable, and she's a real person."[44] That last phrase—"she's a real person"—encapsulates the potential power of parasocial relationships with television scientists to shape perceptions of the profession.

To test for evidence of the "Scully effect" on a wider scale, 21st Century Fox (which made *The X-Files*) and the Geena Davis Institute on Gender in Media sponsored a survey of more than 2,000 women.[45] Their study found that respondents who had regularly watched *The X-Files* were particularly likely to say young women should be encouraged to study STEM and to say they would encourage their own daughter or granddaughter to enter a STEM field. Frequent viewers of the program were also more likely to have studied and worked in STEM fields themselves. Furthermore, almost all respondents who were familiar with Scully's character regarded her as a role model for girls and women. As the study's authors argue, these findings suggest that the show challenged viewers' stereotypes about science and, in doing so, influenced the career possibilities women in its audience envisioned for themselves.

The evidence for a "Scully effect" does share one limitation with our own evidence for cultivation effects and genre-specific viewing effects: it comes from looking at correlations within survey data. However, these effects dovetail with findings from randomized experimental studies showing that television programs can shape viewers' beliefs about a wide range of topics, including perceptions related to gender, race and ethnicity, LGBTQ identities, and disability.[46] The potential effects we've discussed also fit with the results from an experiment we conducted to test how another program, *The Big Bang Theory*, might influence stereotypes of scientists along with perceptions of opportunities for women working in science.

The Big Bang Theory and Stereotypes of Science

Before we get to that experiment, however, let's take a closer look at the show's messages so we can speculate about their impact. Previous studies of *The Big Bang Theory* suggest two salient features of how it portrays its scientist characters. The first is that it tends to depict them in stereotypical, if sympathetic, ways. The show presents all four of the original scientist characters—Leonard, Sheldon, Raj, and Howard—as possessing classic nerd traits such as being socially inept and wearing unfashionable clothes, and it portrays Amy in much the same way.[47] Furthermore, these characters embrace nerd culture, from *Star Trek* to comic books. Sheldon, in particular, reflects the stereotype of the strange and anti-social science nerd. Monika Bednarek's linguistic analysis of his dialogue demonstrates how even his speaking style sets him apart as "a full-blown nerd/geek."[48] For example, he uses academic language in everyday conversations ("I accept your premise, I reject your conclusion") and frequently emphasizes his own superior knowledge ("I don't need validation from lesser minds").[49]

A second key pattern in how *The Big Bang Theory* portrays scientists revolves around its depiction of gender roles. Beyond the contrast between the four original male scientists and the non-scientist Penny, the show's most prominent treatments of gender revolve around Bernadette and Amy. Two separate analyses of the show, one by Heather McIntosh and the other by Margaret Weitekamp, both conclude that its portrayals of these two characters sometimes challenge but ultimately reinforce traditional gender stereotypes and roles. Although Bernadette is a microbiologist, the show rarely presents her in a laboratory setting and often highlights her unprofessional behavior—for example, when she crosses the Ebola virus with the cold virus.[50] Furthermore, she downplays her own intelligence to protect the ego of her boyfriend, Howard. The program does more to emphasize Amy's scientific knowledge and accomplishments in neurobiology, as when she obtains funding for "a brand-new state-of-the-art fMRI machine."[51] In story terms, however, her intelligence also positions her as a compatible romantic partner for Sheldon.[52] As Weitekamp writes, "the female scientists serve primarily to support the character development experienced by the core male actors."[53]

In another study, Rachel Li and Lindy A. Orthia drew on a series of focus group discussions to examine how both scientists and non-scientists interpreted what *The Big Bang Theory* says about science.[54] Several themes emerged in their data. One is how the show depicts the work that goes into the scientific process, from Raj using his telescope to Amy dissecting brains. Another theme is how *The Big Bang Theory* presents science as "subjective and theory-laden"—for example, by showing Leonard and another physicist debating string theory and loop quantum gravity.[55] Yet another theme is how the show presents science as "socially and culturally embedded, with some characters—particularly Sheldon—attaching great importance to distinctions in rank (people with doctorates versus people with Masters degrees) and across fields (scientists versus engineers, physics versus geology)."[56]

Li and Orthia's focus group participants also discussed—and sometimes endorsed—the show's stereotypical portrayals of its scientists. "I hang around physicists a lot," explained one participant, "and they do act a bit strange in a sense that they're very empirical and evidence-driven so they pick a lot of what most would consider 'arguments.' But they are more interested in getting the truth."[57]

A *Big Bang Theory* Experiment

With all this in mind, we conducted an experiment to test how watching *The Big Bang Theory* influenced audience members' stereotypes of scientists and perceptions of gender bias in science. In the spring of 2017, we recruited 437 college undergraduates to take part in our study. These participants were by no means typical of all television viewers. However, they did come from a particularly relevant subset of the public: students who may someday become scientists, media professionals, or both.

Our experimental approach allowed us to isolate the effects of specific television scenes on viewers. We randomly assigned each participant to watch one of five video clips. Four of these came from *The Big Bang Theory*:

- The first video (hereafter, the "Sheldon" clip) features Sheldon attempting to teach Penny physics. She proves to be a poor student. After he says, "MA equals MG, and what do we know from this?" she answers, "Uh, we know that ... Newton was a really smart cookie. Oh! Is that where Fig Newtons come from?"[58] For his part, Sheldon is an arrogant and unsympathetic teacher. When Penny begins to cry and calls herself stupid, he tells her, "That's no reason to cry. One cries because one is sad. For example, I cry because others are stupid, and it makes me sad." This clip depicts the classic stereotype of the scientist as an eccentric and socially awkward man.
- The second video (the "Leonard" clip) shows Penny visiting Leonard at his lab. She makes a series of characteristically ditzy comments about his equipment and then asks him about his current project. He demonstrates a "front-projected holographic display combined with laser-based finger tracking" and says, "The holographic principle suggests that ... our lives are really just acting out a painting on the largest canvas in the universe."[59] She responds

by telling him to take off his clothes. We chose the clip as one that depicts a male scientist being nerdy but socially adept and romantically successful.

• The third video (hereafter, the "Bernadette" clip) features Bernadette worrying about how her employers will react to her pregnancy. "I'm up for a major immunotherapy study," she tells Amy, "and if they find out I'm pregnant they might give it to someone else ... I know they would—they did it to Barbara Chen last year when I told everyone *she* was pregnant."[60] We chose this clip as an example of the show portraying a women experiencing gender bias in a scientific workplace.

• In the fourth video (the "Amy" clip), Amy attends a party at her university with Penny. "To be honest," says Amy, "it's not like a 'party' party, it's more like a gathering where scientists of different disciplines get together to share their work and keep current on what's going on in other fields." The host delivers some geology-related puns and tells Penny that Amy is "the coolest girl on campus."[61] This clip provides an example of the show portraying a woman scientist as popular among her scientific peers—if still geeky by Penny's "normal" standards.

The fifth video came from a different television sitcom (*The New Girl*) and didn't feature any science-related content. Viewers assigned to watch it served as the control group and provided a baseline for comparison to the participants who watched the *Big Bang Theory* clips.

We then asked each participant a series of questions about science and scientists. Three questions captured agreement with common stereotypes of scientists as "odd and peculiar people" who "tend to be socially awkward" and "wear unfashionable clothes." Another pair of questions measured beliefs about whether "women are underrepresented in science" and whether "women who work in science are likely to experience gender bias."

The Effects of Watching *The Big Bang Theory*

Characters on *The Big Bang Theory* often discuss the results of their experiments, which range from laboratory tests involving particle physics to trials at romantic cohabitation to slipping alcohol into someone else's "virgin" mixed drinks (an unethical research practice we don't condone). The results of our experiment *about* the program demonstrate that scenes from it influenced viewers' perceptions in more ways than one.

To begin with, viewing *The Big Bang Theory* led to greater agreement with stereotypes of scientists as strange, socially awkward, and unfashionable (Figure 3.4). Among the participants who watched a clip from the show, 53% agreed that scientists are odd and peculiar, 51% agreed that scientists tend

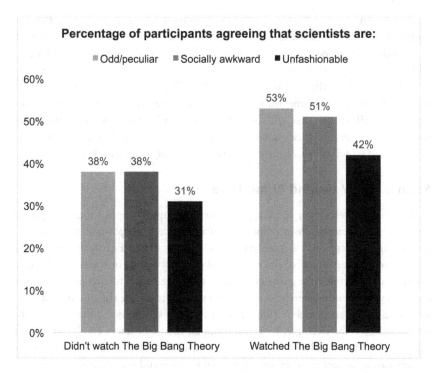

FIGURE 3.4 Stereotypes of scientists, by *Big Bang Theory* watching condition (*The Big Bang Theory* experiment, 2017)

to be socially awkward, and 42% agreed that scientists are unfashionable. By comparison, participants who did *not* watch a clip from *The Big Bang Theory* were 15 percentage points less likely to see scientists as odd and peculiar, 13 points less likely to see scientists as socially awkward, and 11 points less likely to see scientists as unfashionable. Seeing a nerdy scientist on television—whether it was Sheldon, Leonard, Bernadette, or Amy—encouraged viewers to view scientists in general as being geeky.

Meanwhile, the impact of watching *The Big Bang Theory* on perceptions of gender bias depended on which clip participants saw. Seeing the Amy clip led viewers to perceive less gender bias in science (66% agreement, versus 80% in the control condition) and less underrepresentation of women in science (76% agreement, versus 86% in the control). These results suggest that television portrayals of a woman scientist as "cool" may foster more optimistic perceptions of the real-world climate for women in science. Watching the Leonard clip also reduced perceptions of gender bias in science (65% agreement) and underrepresentation of women in science (72% agreement), perhaps because it portrayed Penny's enthusiastic engagement with her romantic partner's scientific research. Neither of the other clips produced any discernible impact on perceptions of gender bias and representation in science.

Taken together, the results of our *Big Bang Theory* experiment point to the power of individual prime time shows to mold perceptions of scientists—for good or for ill. In this case, we found that parasocial contact with sitcom scientists promoted stereotypical views of scientists. We also found that *some* instances of such contact led to rosier views (warranted or not) of gender bias in science.

Science on TV beyond Prime Time

Entertainment television viewing can carry complex implications for public perceptions of science. As a whole, it can reinforce both positive views of scientists and perceptions of scientific work as dangerous. Furthermore, it may shape how viewers perceive the real-world gender balance in science by leading them to overestimate the ratio of women working in fields such as astronomy, chemistry, computer science, and engineering. On top of this, genres of programming such as science fiction can carry their own unique consequences for public beliefs—for example, by reinforcing an image of science as hazardous to one's health. Even individual programs such as *The X-Files* and *The Big Bang Theory* can influence how viewers perceive science and scientists by echoing or challenging common stereotypes.

Note, however, that these conclusions revolve around *fictional* entertainment television. Other sorts of television programs focus on conveying real-world information to viewers. Indeed, entire channels—such as the Discovery Channel, Animal Planet, and the National Geographic Channel—are devoted to shows that blend entertainment and informational values. This genre of "infotainment" merits closer investigation as a potential influence on perceptions of science.

Notes

1 Heyman, Karen, "Talk nerdy to me," *Science* 320, no. 5877 (2008): 740–741; Weitekamp, Margaret A., "The image of scientists in *The Big Bang Theory*," *Physics Today* January (2017): 40–48.

2 Heyman, "Talk nerdy," 741.

3 Wikipedia, "*The Big Bang Theory* U.S. ratings," September 7, 2020, https://en.wikipedia.org/wiki/The_Big_Bang_Theory#U.S._ratings.

4 See the Appendix for details about this survey.

5 Stossel, Scott, "The man who counts the killings," *Atlantic*, May 1997.

6 Gerbner, George, and Larry Gross, "Living with television: The violence profile," *Journal of Communication* 26, no. 2 (1976): 172–199; Gerbner, George, Larry Gross, Michael Morgan, and Nancy Signorielli. "The 'mainstreaming' of America: Violence profile number 11," *Journal of Communication* 30, no. 3 (1980): 10–29.

7 Morgan, Michael, and James Shanahan, "The state of cultivation," *Journal of Broadcasting & Electronic Media* 54, no. 2 (2010): 337–355.

8 Potter, W. James, "A critical analysis of cultivation theory," *Journal of Communication* 64, no. 6 (2014): 1015–1036.

9 Gerbner, George, "Science on television: How it affects public conceptions," *Issues in Science and Technology* 3, no. 3 (1987): 109–115; Gerbner, George, Larry Gross, Michael Morgan, and Nancy Signorielli, "Scientists on the TV screen," *Society* 18, no. 4 (1981): 41–44.

10 Gerbner, "Science on television," 115.

11 Nisbet, Matthew C., Dietram A. Scheufele, James Shanahan, Patricia Moy, Dominique Brossard, and Bruce V. Lewenstein, "Knowledge, reservations, or promise? A media effects model for public perceptions of science and technology," *Communication Research* 29, no. 5 (2002): 584–608.

12 Nisbet et al., "Knowledge, reservations, or promise?" 587.

13 Nisbet et al., "Knowledge, reservations, or promise?" 588.

14 Dudo, Anthony, Dominique Brossard, James Shanahan, Dietram A. Scheufele, Michael Morgan, and Nancy Signorielli, "Science on television in the 21st century: Recent trends in portrayals and their contributions to public attitudes toward science," *Communication Research* 38, no. 6 (2011): 754–777.

15 Kirby, David A., "The changing popular images of science," in *The Oxford handbook of the science of science communication*, ed. Kathleen Hall Jamieson, Dan Kahan, and Dietram A. Scheufele, Oxford University Press, 2017: 291–300; Nisbet, Matthew C., and Anthony Dudo, "Entertainment media portrayals and their effects on the public understanding of science," in *Hollywood chemistry: When science met entertainment*, ed. Donna J. Nelson, Kevin R. Grazier, Jaime Paglia, and Sidney Perkowitz, American Chemical Society, 2013: 241–249; Perkowitz, Sidney, *Hollywood science: Movies, science, and the end of the world*, Columbia University Press, 2007.

16 We are grateful to Nancy Signorielli for sharing these findings with us. For more about the Cultural Indicators Project, see Signorielli, Nancy, Michael Morgan, and James Shanahan, "The violence profile: Five decades of cultural indicators research," *Mass Communication and Society* 22, no. 1 (2019): 1–28.

17 See the Appendix for details about this survey.

18 Johnson-Smith, Jan, *American science fiction TV: Star Trek, Stargate, and beyond*, Wesleyan University Press, 2005; Telotte, Jay P., ed., *The essential science fiction television reader*, University Press of Kentucky, 2008.

19 Nisbet, Matthew C., and Robert K. Goidel, "Understanding citizen perceptions of science controversy: Bridging the ethnographic—survey research divide," *Public Understanding of science* 16, no. 4 (2007): 421–440.

20 Brossard, Dominique, and Anthony Dudo, "Cultivation of attitudes toward science," in *Living with television now: Advances in cultivation theory and research*, ed. Michael Morgan, James Shanahan, and Nancy Signorielli, Peter Lang, 2012: 120–146.

21 The Oxford English Dictionary defines "biatch" as a "disparaging form of address."

22 Orthia, Lindy A., and Rachel Morgain, "The gendered culture of scientific competence: A study of scientist characters in *Doctor Who* 1963–2013," *Sex Roles* 75, no. 3–4 (2016): 79–94.

23 Morgain, Rachel, and Lindy Orthia, "Ahead of its time: *Doctor Who*'s 56 inspiring female scientists," *The Conversation*, May 18, 2016, https://theconversation.com/ahead-of-its-time-doctor-whos-56-inspiring-female-scientists-58491.

24 See Chapter 1.

25 National Science Board, "The state of U.S. science and engineering 2020," National Science Foundation/National Science Board, 2020, https://ncses.nsf.gov/pubs/nsb20201.

26 *The Big Bang Theory*, "The Transporter Malfunction" (Season 5, Episode 20).

27 See Chapter 1.

28 Harbour, Wendy. "*The Big Bang Theory*: Mad geniuses and the freak show of higher education," *Review of Disability Studies: An International Journal* 11, no. 2 (2015); Walters, Shannon, "Cool aspie humor: Cognitive difference and Kenneth Burke's comic corrective in *The Big Bang Theory* and *Community*," *Journal of Literary & Cultural Disability Studies* 7, no. 3 (2013): 271–288.

29 Collins, Paul, "Must geek TV," *Slate*, August 30, 2010, https://slate.com/news-and-politics/2010/08/an-emmy-for-the-first-sitcom-character-with-asperger-s-the-big-bang-theory-s-sheldon-cooper.html.

30 *The Big Bang Theory*, "The Closure Alternative" (Season 6, Episode 21).

31 GLAAD, *Where we are on TV, '17–'18: GLAAD's annual report on TV inclusion*, 2017, https://glaad.org/files/WWAT/WWAT_GLAAD_2016-2017.pdf; Squire, Tari Hartman, Kristina Kopić, and Daryl "Chill" Mitchell, *The Ruderman white paper on the challenge to create more authentic disability casting and representation on* TV, Ruderman Family Foundation, 2017, https://rudermanfoundation.org/wp-content/uploads/2017/09/tv_challenge_wp.pdf.

32 GLAAD, *Where we are*.

33 Morgan, Michael, James Shanahan, and Nancy Signorielli, eds., *Living with television now: Advances in cultivation theory and research*, Peter Lang, 2012.

34 Farnall, Olan, and Kim A. Smith., "Reactions to people with disabilities: Personal contact versus viewing of specific media portrayals," *Journalism & Mass Communication Quarterly* 76, no. 4 (1999): 659–672; Jones, Philip Edward, Paul R. Brewer, Dannagal G. Young, Jennifer L. Lambe, and Lindsay H. Hoffman, "Explaining public opinion toward transgender people, rights, and candidates," *Public Opinion Quarterly* 82, no. 2 (2018): 252–278; Morgan et al., *Living with television*.

35 See the Appendix for details about this survey.

36 National Science Board, "The state of U.S. science and engineering 2018," National Science Foundation/National Science Board, 2018, https://www.nsf.gov/statistics/2018/nsb20181/report, Appendix Table 3–12.

37 Li, Pei-ying Rashel, "Communicating science through entertainment television: How the sitcom *The Big Bang Theory* influences audience perceptions of science and scientists," unpublished doctoral dissertation, Australian National University, May 2016.

38 SMZ, "Asian Stereotypes in *The Big Bang Theory* and *The Boondocks*: The inability of television to multitask with subalterns," *Radical Compounds*, Feb. 18, 2013, https://radicalcompounds.wordpress.com/2013/02/18/asians-in-the-big-bang-theory-and-the-boondocks.

39 Giles, David C., "Parasocial interaction: A review of the literature and a model for future research," *Media Psychology* 4, no. 3 (2002): 279–305; Perse, Elizabeth M., and Rebecca B. Rubin, "Attribution in social and parasocial relationships," *Communication Research* 16, no. 1 (1989): 59–77.

40 Schiappa, Edward, Peter B. Gregg, and Dean E. Hewes, "Can one TV show make a difference? *Will & Grace* and the parasocial contact hypothesis," *Journal of Homosexuality* 51, no. 4 (2006): 15–37.

41 Hoffner, Cynthia A., and Elizabeth L. Cohen, "Responses to obsessive compulsive disorder on Monk among series fans: Parasocial relations, presumed media influence, and behavioral outcomes," *Journal of Broadcasting & Electronic Media* 56, no. 4 (2012): 650–668.

42 Vineyard, Jennifer, "Nearly everything *The X-Files'* David Duchovny and Gillian Anderson said this weekend," *Vulture*, Oct. 14, 2013, www.vulture.com/2013/10/david-duchovny-gillian-anderson-nycc-paley-center-quotes.html.

43 Lane, Carly, "The new *X-Files* science advisor explains how the reboot will stay 'realistic,'" *Vice*, Aug. 10, 2015, https://motherboard.vice.com/en_us/article/nzeppk/the-new-x-files-science-advisor-explains-how-the-reboot-will-stay-realistic.

44 Lane, Carly, "The new *X-Files*."

45 21st Century Fox, Geena Davis Institute on Gender in Media, and J. Walter Thompson Intelligence, "The 'Scully effect': I want to believe … in STEM," Geena Davis Institute on Gender in Media, 2019, https://seejane.org/wp-content/uploads/x-files-scully-effect-report-geena-davis-institute.pdf.

46 See, e.g., Elliott, Timothy R., and E. Keith Byrd, "Attitude change toward disability through television: Portrayal with male college students," *International Journal of Rehabilitation Research* 7, no. 3 (1984): 320–322; Hall, Heather, and Patricia Minnes, "Attitudes toward persons with Down syndrome: The impact of television," *Journal of Developmental and Physical Disabilities* 11, no. 1 (1999): 61–76; Mastro, Dana, and Riva Tukachinsky, "The influence of exemplar versus prototype-based media primes on racial/ethnic evaluations," *Journal of Communication* 61, no. 5 (2011): 916–937; Schiappa, Edward, Peter B. Gregg, and Dean E. Hewes, "The parasocial contact hypothesis," *Communication Monographs* 72, no. 1 (2005): 92–115; Ward, L. Monique, and Kristen Harrison, "The impact of media use on girls' beliefs about gender roles, their bodies, and sexual relationships: A research synthesis," in *Featuring females: Feminist analyses of media*, ed. Ellen Cole and Jessica Henderson Daniel, American Psychological Association, 2005, 3–23.

47 Bednarek, Monika, "Constructing 'nerdiness': Characterisation in *The Big Bang Theory*," *Multilingua* 31, no. 2 (2012): 199–229; Cooper, W. Jeff, "Stereotypes in television and film: The impact of *The Big Bang Theory*," *Men & Masculinities Knowledge Community* Fall (2014): 8–10; McIntosh, Heather, "Representations of female scientists in *The Big Bang Theory*," *Journal of Popular Film and Television* 42, no. 4 (2014): 195–204; Weitekamp, Margaret A. "'We're physicists': Gender, genre and the image of scientists in *The Big Bang Theory*," *Journal of Popular Television* 3, no. 1 (2015): 75–92.

48 Bednarek, Monika, "Constructing 'nerdiness,'" 223.

49 Bednarek, Monika, "Constructing 'nerdiness,'" 208–211.

50 McIntosh, "Representations of female scientists," 198.

51 *The Big Bang Theory*, "The Retraction Reaction" (Season 11 Episode 2).

52 McIntosh, "Representations of female scientists."

53 Weitekamp, "'We're physicists,'" 86.

54 Li, Rashel, and Lindy A. Orthia, "Communicating the nature of science through *The Big Bang Theory*: Evidence from a focus group study," *International Journal of Science Education, Part B* 6, no. 2 (2016): 115–136.

55 Li and Orthia, "Communicating the nature," 116.

56 Li and Orthia, "Communicating the nature," 116.

57 Li and Orthia, "Communicating the nature," 129.

58 *The Big Bang Theory*, "The Gorilla Experiment" (Episode 10, Season 3).

59 *The Big Bang Theory*, "The Holographic Excitation" (Episode 5, Season 6).

60 *The Big Bang Theory*, "The Military Miniaturization" (Episode 2, Season 10).

61 *The Big Bang Theory*, "The Dependence Transcendence," (Episode 3, Season 10).

4

DOCUMENTARY SCIENCE

Neil deGrasse Tyson: This adventure is made possible by generations of searchers strictly adhering to a simple set of rules. Test ideas by experiments and observations. Build on those ideas that pass the test. Reject the ones that fail. Follow the evidence wherever it leads, and question everything. Accept these terms, and the cosmos is yours.

—*Cosmos: A Spacetime Odyssey* (2014; Episode 1:
Standing Up in the Milky Way)

Adam Savage: Remember, kids, the only difference between screwing around and science is writing it down.

—*MythBusters* (2012; Season 10,
Episode 8: The Bouncing Bullet)

"The cosmos is all that is, or ever was, or ever will be," astronomer Carl Sagan proclaims at the beginning of his 1980 documentary television series, *Cosmos: A Personal Voyage*. "We're going to explore the cosmos in a ship of the imagination, unfettered by ordinary limits on speed and size." Over the course of 13 episodes, he covers topics ranging from the origins of the universe to the potential fate of humanity. The show blends this scientific information with elements designed to engage and entertain viewers, including special effects that were state-of-the-art for their time, soaring synthesizer music by popular musician Vangelis, and Sagan's own starry-eyed, poetic narration. "The cosmos is also within us," he tells his audience, in one typical flight of language. "We're made of star-stuff."

DOI: 10.4324/9781003190721-4

Cosmos was a ratings hit for the Public Broadcasting Service (PBS), drawing more viewers than any US public television program had before it.[1] Sagan himself became the most famous scientist of his generation; he was so recognizable that *Tonight Show* host Johnny Carson took to imitating the astronomer's fondness for the word "billions."[2] Not that Sagan seemed to mind; he frequently appeared on the late-night comedian's program to discuss scientific topics. He also wrote a science fiction novel about the discovery of extraterrestrial intelligence, *Contact*, that was later adapted as a successful film starring Jodie Foster. Many younger scientists and science communicators—including astrophysicist Neil deGrasse Tyson and educational television host Bill Nye—would later credit him for helping to inspire their own careers.[3]

Five years after the broadcast of *Cosmos*, the newly formed Discovery Communications launched a documentary-themed cable television channel in the hopes of drawing an audience that would attract science and technology advertisers.[4] Unlike PBS, this network didn't depend on public funding or donations. Unlike the major broadcast networks, it didn't try to appeal to a general audience, either. Instead, the new economics of the cable industry allowed The Discovery Channel to flourish by building a smaller but loyal audience for its documentary programs.[5] As it evolved from purchasing second-hand content to developing its own programs, the channel and its parent company—which eventually rebranded itself as Discovery Inc.—found ratings success through shows that combined scientific information with reality television-style production and dramatic appeal. Over the years, the network's biggest hits have included *Shark Week*, an annual block of shark documentaries, and the trial-and-error based *MythBusters*.

Nor is The Discovery Channel the only television network to build viewership through science documentary programming. Prompted by The Discovery Channel's popularity, other cable channels—including Animal Planet and the National Geographic Channel—have developed numerous science-themed shows. *Cosmos* itself underwent a revival in 2014, when Fox Network and the National Geographic Channel teamed up to produce a new version hosted by Neil deGrasse Tyson. During all of this, PBS has continued to air its own science documentary programs, including *NOVA* and *Nature*.

Collectively, such science documentary programs reach a sizable audience. Almost half the respondents (45%) in a 2017 Pew Research Center survey of the US public said they regularly got science news from "documentaries or other science video programs."[6] Among other nonfiction media forms, only general news outlets ranked higher as frequent sources of science news. Moreover, the people who watched science documentaries and videos placed substantial faith in them, with two-thirds (68%) saying these sources "get the facts right when it comes to science" most of the time.

Beneath their broad aura of credibility, science documentary programs vary considerably in how they present science. In particular, they differ in how they combine *informational* values and *entertainment* values. Virtually all science-themed television shows reflect both sets of values to at least some degree; as a result, some observers have labeled this type of programming "infotainment" or "edutainment."[7] Yet specific documentary programs can fall anywhere on a continuum between the purely informational and the purely entertaining. At one end of the spectrum, *NOVA* tends to emphasize educational content, as do both versions of *Cosmos*. At the other end, The Discovery Channel's *Shark Week* and its "docufiction" specials such as *Mermaids: The Body Found* often highlight drama at the expense of informational value. The same network's *MythBusters* falls somewhere in between these two poles with its mixture of scientific hypothesis-testing and television-friendly explosions.

Depending on their subject matter, as well as the ways they balance information versus entertainment, science documentary programs may influence public perceptions in a variety of directions. On the one hand, watching sensationalized documentaries and docufictions could fuel distorted understandings of nature, prime public fears of the natural world, and erode trust in scientific institutions. On the other hand, viewing documentary shows such as *Cosmos*, *NOVA*, and *MythBusters* may help dispel stereotypes of scientists and reshape perceptions of scientific work. Such effects, in turn, could ultimately spill over to audience members' decisions about whether to support scientific research, engage in everyday scientific activities, or even follow the likes of Carl Sagan and Neil deGrasse Tyson in pursuing scientific careers.

Science as Entertainment on Cable Television: Drama and Docufictions

One company, Discovery Inc., has dominated science-themed documentary cable television since the 1980s. As of the late 2010s, it operated 20 channels and accounted for a fifth of all US cable viewership.[8] Along with its original network, The Discovery Channel, the company owns two other channels with documentary science content: Animal Planet and Science, both launched in 1996. Taken together, these three channels have featured a wide variety of science-related shows. In addition to *Shark Week* and *MythBusters*, The Discovery Channel has run programs such as *Expedition Unknown* (about legends and mysteries involving archaeology), *Invisible Killers* (about viruses), *Tesla's Death Ray* (about the famed inventor's claim to have designed a lethal energy beam), and *Man-Eating Python* (self-explanatory). Animal Planet's most popular shows have included *Crocodile Hunter* (hosted by Steve Irwin until his death in 2006), *Whale Wars* (about conservationists trying to stop whalers), and *River Monsters* (about freshwater predators). The lower-profile Science has featured original series such as *Outrageous Acts of Science*, in which science and engineering experts discuss popular internet videos, alongside reruns of Discovery Channel content.

In 2001, the National Geographic Society and Fox Cable Networks launched their own competitor in the realm of documentary television: the National Geographic Channel. This channel has co-produced the new *Cosmos* series along with programs running the gamut from *StarTalk* (also hosted by Neil deGrasse Tyson) and *One Strange Rock* (about the science of the earth) to *Rocket City Rednecks* (an engineering-themed program featuring moonshine-powered rockets and a "hillbilly hovercraft"). The channel became part of Walt Disney Television following a 2017 media merger.

In looking at science on cable television, it's important to remember that The Discovery Channel, Animal Planet, Science, and the National Geographic Channel are commercial networks that depend on cable subscriptions and advertising revenue to generate profits. Accordingly, their content reflects entertainment values along with—and sometimes in preference to—informational values.[9] These multiple and potentially contradictory priorities can carry mixed implications when it comes to public understandings of science.

In particular, science documentary shows often emphasize stock characters and storylines, vivid metaphors, and striking visuals to draw viewers.[10] For example, David Pierson found that The Discovery Channel's nature programming tends to anthropomorphize animals in conventional plots revolving around good and evil.[11] Some animals on shows such as *Wild Discovery* play the role of heroes, including a mother bear tending to her cubs and a young bear learning to catch salmon. Others, such as a troop of baboons, play different roles: "criminals" who steal human food or "victims" of medical testing.[12] Not surprisingly, "cute" animals (such as bears and whales) tend to receive sympathetic portrayals whereas "repulsive" ones (such as snakes) serve to inspire fear.[13] Although these narratives may promote viewers' interest in and connection to science, they also encourage audience members to see nature in dramatized and oversimplified ways—as well as to project their own moral frameworks onto it.

Similarly, Vincent Campbell describes how cable documentary shows use both Hollywood-style narratives and computer-generated imagery (CGI) to enhance their entertainment value.[14] For example, programs such as *Supervolcano* and *Superstorm* rely on the same disaster movie tropes that viewers might see in *The Day After Tomorrow* or *Twister*, including depictions of mass destruction and survivors. In terms of CGI imagery, cable documentaries offer viewers "perceptually realistic" spectacles by recreating dinosaurs (as in *Walking with Dinosaurs*), supernovae (as in *Wonders of the Universe*), and natural disasters (as in *Super Comet* and *Supervolcano*).[15] Some observers have criticized these techniques as sensationalistic, but Campbell argues that their power to resonate with viewers' own experiences and emotions can help foster engagement with the "wonder" of science.[16]

Although scholars have highlighted both the positive and negative implications of "edutainment" in cable documentaries, some shows have stirred strong condemnation from the scientific community for the particular ways they prioritize sensationalism over established scientific practices, knowledge, and ethical principles.[17] In 2014, The Discovery Channel faced a backlash from both herpetologists and viewers for *Eaten Alive*, which network promotions suggested would feature a man allowing an anaconda to swallow and regurgitate him (the snake rejected its meal).[18] One of the special's fiercest critics, evolutionary biologist Christie Wilcox, also wrote an exposé of the same channel's 2016 documentary series *Venom Hunters* that pointed out faulty snake-handling practices on the part of its supposed experts and apparent ethical lapses on the part of its producers.[19] Wilcox criticizes *Eaten Alive* for perpetuating exaggerated fears of human-eating snakes and *Venom Hunters* for encouraging amateur copycats to endanger both themselves and snakes.

Marine scientists responded with similar outrage to a series of aquatic-themed specials on The Discovery Channel and Animal Planet. *Mermaids: The Body Found*, which ran on both channels in 2012, suggests that the US National Oceanographic and Atmospheric Administration (NOAA) had concealed evidence of real-life mermaids. The agency issued a rebuttal, but the program drew high enough ratings that it spawned a 2013 sequel, *Mermaids: New Evidence*.[20] Scientists have criticized such hype-driven shows as potentially damaging to public support for scientific endeavors and trust in scientific agencies. Deep-sea ecologist David Andrew Thaler writes:

> These kinds of programs muddy the waters of education-based television … In the numerous cases of animal abuse, they cause active harm to the wildlife about which they are ostensibly attempting to educate the public. And the bold and outright fabrications of shows like *Mermaids* erode the public's trust in government and scientific organizations. By framing the villain in these productions as real, often nonpartisan, institutions like NOAA, they don't just direct resources away from the agency's actual work by forcing it to respond to a phony controversy; they lend weight to other campaigns aimed at discrediting these organizations.[21]

He even quotes a schoolteacher he met on an airplane trip who told him, "If NOAA is lying to us about the existence of mermaids then they're definitely lying to us about climate change."[22] In short, shows such *Mermaids* can sow distrust in science and scientific organizations by using documentary techniques that create an aura of perceptual realism and build on The Discovery Channel's own reputation for credibility.

Shark Week: Will You Ever Go in the Water Again?

If Thaler and other Discovery Inc. critics are right about the power of sensation-alized documentary shows to promote fears of the natural world and cynicism about scientific organizations, then this type of programming may also carry real-world consequences for scientific initiatives such as conservation efforts. Take *Shark Week*, The Discovery Channel's crown jewel since 1988. The website for the show bills it as "television's longest-running and eagerly awaited summer TV event, delivering all-new groundbreaking shark stories and incorporating innovative research technology."[23] *Shark Week* receives heavy promotion from the channel and regularly draws high ratings.[24] A third of the respondents (34%) in our own October 2016 survey of the US public had seen it, making it in all likelihood the most prominent media portrayal of sharks since the 1975 block-buster film *Jaws* (which was advertised with the tagline, "You'll never go in the water again").[25]

Drawing on a content analysis of episodes from 2001 to 2012, Suzanne Evans examined what frames, or story elements, *Shark Week* used to portray its star predators.[26] She found that entertainment frames dominated, consistent with the cable network's ever-present need to draw viewers. The episodes she coded often focused on shark attacks by featuring interviews with victims and shots of bloody water or struggling swimmers. The series also used reenactments, scary music, and sped-up or slowed-down footage to emphasize such attacks. As Evans points out, *Shark Week's* portrayal of sharks as "killers" exaggerates the threat they pose to humans—and, in doing so, undermines public support for scientists' efforts to preserve endangered species of sharks. In particular, the show triggers fears that can reduce viewers' willingness to help conservation efforts through actions such as discussing the topic with family or friends, signing petitions to public officials, and donating money to conservation organizations.[27]

Evans did find that some aspects of framing on *Shark Week* shifted after 2010, when The Discovery Channel promised to address criticisms from scientists and conservationist organizations. For example, the program included more scientific sources and conservation messages as well as fewer portrayals of real shark attacks. Not long after Evans completed her study, however, the channel returned to its old habit of sensationalizing sharks in a series of docufiction specials. As part of 2013's *Shark Week* block, The Discovery Channel ran *Megalodon: The Monster Shark Lives*, which falsely suggests that an extinct species of shark caused the sinking of a fishing boat. The special was a hit, earning a sequel in 2014 as well as a successor in 2015's *Shark of Darkness: Wrath of Submarine*, which spuriously blames a great white shark for the capsizing of a passenger boat. The network gave viewers a new bait-and-switch in 2017 by promising to show a race between Olympic gold medalist swimmer Michael Phelps and a great white, then delivering a CGI shark (which "defeated" Phelps but didn't eat him).[28] All of this goes to show how difficult it is for commercial cable networks to resist the lure of sensationalism when it draws ratings and advertising revenue.

Building on the finding that *Shark Week* often emphasizes dramatic and violent content, Evans and Jessica Gall Myrick conducted a randomized experiment to test how participants would respond to watching shark attack videos from the program.[29] As it turns out, seeing such videos fanned audience members' anxieties about sharks. The study also found a link between recent viewing of *Shark Week* and fear of sharks, but no link between lifetime viewing of the program and such fear. As Myrick and Evans point out, these results suggest that *Shark Week* influences viewers primarily by activating existing memories—an effect known as *priming*. By contrast, the researchers found no evidence that long-term exposure to the show *cultivated* fear of sharks. In broader terms, the authors' findings highlight the potential for *Shark Week* to hinder real-world conservation efforts by reinforcing the image of sharks as dangerous.

The ominous music included in shark documentaries may further magnify hostile attitudes toward sharks. Just as *Jaws* relies on a famous two-note theme to build suspense, *Shark Week* and other documentary television shows about sharks often feature menacing soundtracks. To capture the impact of these musical choices, Andrew Nosal and his colleagues conducted a series of experiments in which they randomly assigned participants to watch a clip from a shark documentary (the BBC series *Blue Planet: Seas of Life*, which ran on The Discovery Channel in the United States) that included either uplifting music, ominous music, or no music.[30] Afterward, the participants who watched the clip with scary music expressed particularly negative views toward sharks. As the researchers conclude, "an ominous soundtrack may enhance [the] entertainment aspect" of shark documentaries, but "may also impede legitimate shark conservation efforts by biasing viewers' perceptions."[31]

On the positive side, communication scholars and marine scientists have explored ways to counteract the potentially damaging messages in programs such as *Shark Week*. As part of their experiment, Myrick and Evans tested the effects of public service announcements (PSAs) for shark conservation that The Discovery Channel itself ran. These ads included messages such as "We shouldn't be scared *of* sharks, we should be scared *for* them."[32] Myrick and Evans found mixed results: watching the PSAs increased viewers' support for shark conservation but didn't reduce their fear of sharks.

Andrew David Thaler and David Shiffman took a different strategy in responding to Discovery, Inc.'s sensationalized shark docufictions. To counter the misinformation in *Megalodon* and *Shark of Darkness*, these two scientists wrote articles debunking the programs, published them on a popular marine science and conservation website (Southern Fried Science), and then used search-engine optimization techniques and social media to maximize their visibility when the programs ran.[33] Thaler and Shiffman's critique of *Megalodon* drew a relatively small online audience but laid the groundwork for their next effort. When they launched their campaign against *Shark of Darkness*, they succeeded in drawing half a million unique visitors to their website and making their article about the "fake documentary" the top Google search result for "shark of darkness." In addition, they may have helped drive negative tweets about the series. Thaler and Shiffman see the response their campaign generated as a victory for marine conservation efforts over misinformation. Their efforts also illustrate how science communicators can use *transmedia* strategies that cross different media platforms, from search engines to Facebook and Twitter, to provide viewers with tools for critically engaging televised messages about science and placing such messages in broader social contexts.[34]

MythBusters: Busting Stereotypes of Science?

Though *Shark Week* illustrates the power of cable documentary channels to fuel fears of the natural world and undermine scientific efforts, another Discovery Channel program may represent the potential for such channels to reshape public understandings of scientific practices and culture. Whereas the network's popular shark-themed programming tends to emphasize sensationalized drama, its hit series *MythBusters* balances entertainment values and educational values in depicting how science works, what scientific work is like, and who can participate in it.

Launched by The Discovery Channel and an Australian television channel in 2003, *MythBusters* features special effects experts Jamie Hyneman and Adam Savage using demonstrations to test the accuracy of popular myths and legends.[35] In most episodes, a trio of co-hosts—Tory Belleci, Kari Byron, and Grant Imahara—join in their efforts. At the end of each segment, the hosts rule whether the myth is "confirmed," "plausible," or "busted." For example, they examine whether sticking a finger in the barrel of a gun can cause it to backfire (busted), whether it's possible to beat police speed cameras (plausible but difficult), and whether the shockwave from a bullet hitting water can kill fish (confirmed). *MythBusters* ran for 14 seasons before ending in 2016. The following year, Discovery, Inc. rebooted the program with new hosts on another channel, Science.

During its initial run, *MythBusters* often drew large audiences by cable standards. For example, a 2010 episode featuring President Barack Obama drew more than 2 million viewers.[36] Indeed, more than half of the respondents (56%) in our October 2016 survey of the US public said they'd seen the program. *MythBuster* also received critical acclaim, including eight Emmy nominations.[37] Hyneman and Savage themselves became prominent ambassadors for science and engineering, appearing as guests on television talk programs and speaking at teachers' conferences.[38] When the original program ended, Obama taped a message thanking its hosts for "inspiring so many of our young people to ask the big questions about our world, and to seek the answers through math, science, and engineering."[39]

Some scientists and engineers have joined him in praising *MythBusters*. For example, biomedical engineer Erik Zavrel suggests that its portrayals of science and engineering can help viewers—especially young ones, but also adults— learn about how these fields work in the real world.[40] In particular, he argues that the program presents a fundamentally sound take on many key elements of the scientific method, including the formation of hypotheses, the use of control conditions in experiments, the testing of alternative hypotheses, the development of quantifiable measures, and the importance of replication. In doing so, *MythBusters* relies on a range of pedagogical techniques, such as fostering active learning, avoiding technical jargon, using repetition to enhance retention of information, drawing from familiar pop culture sources, and modeling enthusiasm for scientific inquiry. Most of all, the program uses entertaining demonstrations—including frequent explosions—to illustrate its points. In a companion study, Zavrel and Eric Sharpsteen found that high school students who completed a *MythBusters*-based class activity came away with a greater appreciation for experimental methods along with greater confidence in identifying such methods.[41]

At the same time, skeptical commentators and science and technology studies (STS) scholars have highlighted the potential limitations of *MythBusters* when it comes to fostering critical engagement with science. For example, Brian Dunning argues that the show "never truly challenges its audience" in that it generally avoids "busting" widely held pseudoscientific beliefs about topics such as psychics and therapeutic energy fields.[42] Meanwhile, David Kirby suggests that *MythBusters* conveys an "oversimplification of scientific practice" along with "an overly authoritative representation of scientific authority."[43] Though these aspects of the program may discourage critical thought and reinforce deference rather than active engagement, Dunning and Kirby also note how *MythBusters* provides viewers with the tools and self-efficacy to participate in science. The former lauds the show for using scientific tests to answer questions, and the latter cites the program as helping to "demystify" science.[44]

Echoing the latter point, Zavrel argues that *MythBuster* not only portrays the *process* of science but also offers an accessible depiction of its *culture*, including its social aspects.[45] Unlike the stereotypical mad scientists of old Hollywood movies, the hosts of *MythBusters* come across as down-to-earth, funny, and likable people. They also go out into the world to work with experts ranging from police officers to medical doctors, conveying an image that dovetails with STS-based conceptions of collaborative and participatory science.[46] Even Hyneman and Savage's own background as special effects experts rather than traditional scientists illustrates how laypeople can play an active role in scientific research and science communication—a key principle behind the *public understanding of science model*, which emphasizes the contributions and insights that everyday citizens can make to scientific knowledge, methods, and practice.[47]

Furthermore, *MythBusters* offers viewers a vision of science that features both careful practices and principled inquiry. Hyneman and Savage's demonstrations are often potentially dangerous, but the program highlights their safety precautions, including their use of protective suits and masks. The hosts also exhibit ethical behavior (for example, by using fake fish rather than real ones in demonstrating the effects of shooting fish in a barrel) and address the social impact of science (as when they test how viruses spread).[48]

Given all this, watching *MythBuster* could lead viewers to perceive science as an enjoyable profession where people work together in a responsible way, thereby helping to dispel stereotypes of scientific work as socially isolated, scary, and generally unappealing. In true *MythBusters* spirit, we'll revisit this hypothesis with our own data to test whether it holds up. Before we do so, however, let's look at two programs that tilt even further toward the informational end of the documentary television spectrum: *Cosmos* and *NOVA*.

Neil deGrasse Tyson and the *Cosmos* Revival

Although programs such as *Shark Week* and *MythBusters* helped make The Discovery Channel the most prominent documentary cable channel of the 2010s, two other networks—Fox and National Geographic—produced one of the decade's highest-profile documentary television programs about science: *Cosmos: A SpaceTime Odyssey*. Unlike the original *Cosmos*, the new version aired, not on PBS, but on commercial broadcast and cable networks. As such, its making, content, and distribution reflected both the educational aspirations of the original series and the ratings considerations that drive for-profit television—including the need for a star scientist to follow in Carl Sagan's footsteps.[49]

The revival of *Cosmos* began with a meeting between Neil deGrasse Tyson, director of New York's Hayden Planetarium, and Seth MacFarlane, creator of the Fox animated television series *Family Guy*. The astrophysicist and the Hollywood writer-producer discovered they had two things in common: an interest in fostering public engagement with science and an admiration for Sagan's original *Cosmos*.[50] Working with Sagan's co-writer and widow, Ann Druyan, they planned a new version of the series that Tyson would host. "There has never been a more important time for *Cosmos* to re-emerge than right now," MacFarlane said when the series premiered in 2014, "because of the fact that we have in too many ways roundly ignored and rejected science when it used to be a source of pride for the country, and for the species."[51]

Like its predecessor, *Cosmos: A SpaceTime Odyssey* features scientific information conveyed through engaging narration and elaborate special effects. Reflecting the involvement of MacFarlane, the reboot also includes animated sequences depicting the history of scientific research. The first episode concludes with a tribute in which Tyson describes his experience, as a 17-year-old high school student, of receiving an invitation from Sagan to visit his lab at Cornell University. Tyson accepted and rode a bus from his home in the Bronx to Ithaca, New York. "I already knew I wanted to become a scientist," he tells viewers, "but that afternoon I learned from Carl the kind of person I wanted to become. He reached out to me and to countless others, inspiring so many of us to study, teach, and do science. Science is a cooperative enterprise, spanning the generations, a community of minds reaching back to antiquity and forward to the stars."

Tyson's *Cosmos* covers some of the same ground as Sagan's series—including the history of the universe, the formation of the solar system, and the development of life on earth and maybe elsewhere—while incorporating new scientific insights and addressing new topics of concern, such as climate change. In doing so, the program highlights the importance of the scientific process along with the lives of individual scientists. Sometimes it illustrates how scientific work can be a dangerous and solitary enterprise, conducted by people at odds with the rest of society—as in the case of Giordano Bruno, the 16th-century Italian astronomer burned at the stake for challenging the notion that the Earth is the center of the universe. Much more often, however, *Cosmos: A SpaceTime Odyssey* depicts science as a collaborative enterprise. For example, it portrays how Isaac Newton and Edmund Halley (of Halley's Comet fame) partnered to study the laws of the universe, how Clair Patterson and Harrison Brown worked together to identify the age of the Earth, and how a team of researchers at Harvard Computers—including Annie Jump Cannon, Henrietta Swan Leavitt, and Cecilia Payne—classified the galaxy's stars.

All told, an estimated 45 million US viewers watched at least part of *Cosmos* in 2014.[52] Two years later, our October 2016 survey found that more than a quarter of the US public (28%) had seen it. The series received critical acclaim, as well, including four Emmy Awards.[53] A new season, *Cosmos: Possible Worlds*, premiered in 2020.[54]

The *Cosmos* revival cemented its host's status as one of the most prominent science popularizers of his time, the same role played by Sagan a generation before. Prior to 2014, Tyson had already achieved a high level of visibility through his writings, his media appearances (including frequent visits to Stephen Colbert's *The Colbert Report* and *The Late Show*), and his role in demoting Pluto from the status of planet.[55] Tyson achieved even greater fame following the first season of *Cosmos: A SpaceTime Odyssey*. In fact, he became such a celebrity that comedians

Keegan-Michael Peele and Jordan Key spoofed him on their television program, *Key & Peele*, in 2015 (just as Johnny Carson had once imitated Carl Sagan on *The Tonight Show*). Their sketch features a Tyson who dodges arguments with his wife by launching into narration about the marvels of the universe. For example, when she asks him why he can't keep track of little details like walking the dog, he tells her, "Well, actually, it's the little details that cannot be kept track of, by definition. In 1927, a German university lecturer named Werner Heisenberg came to a seemingly paradoxical conclusion: that the more we know about the position of a particle in physical space, the less we know about its momentum, and vice versa."

Our poll results demonstrate how much of a scientific icon Tyson had become in the wake of *Cosmos*. Most of the respondents in our July 2016 survey of the US public were familiar enough with Tyson to hold an opinion about him, and most of those who rated him held a positive opinion.[56] When we asked respondents in our October 2016 survey of the US public who came to mind when they thought about scientists on TV shows or in movies, almost one in ten (9%) mentioned Tyson, making him the second most-named real-world figure after Bill Nye the Science Guy. Such findings suggest that Tyson is one of the most visible and popular science communicators in the media today. If so, then perhaps his show has fostered more positive views of scientists, encouraged viewers to engage with science, and even provided a role model for potential scientists to emulate—just as Sagan's program did for Tyson.

Yet the case of *Cosmos* also illustrates the challenges in using a documentary television platform to speak to the public about science. In a 2014 survey of the US public, Heather Akin and her colleagues found that respondents who had watched at least one episode of *Cosmos: A SpaceTime Odyssey* already held relatively high levels of science interest and knowledge, making them the "proverbial choir" for its messages.[57] These results highlight an important point to keep in mind when weighing the potential effects of science documentary programs that prioritize information over entertainment values: they may fail to reach members of the public with the lowest levels of curiosity about science.[58] On top of this, the "missing audience" hypothesis outlined by researchers such as Dan Kahan suggests that some potential viewers avoid the likes of *Cosmos* because they perceive such programs as threatening to their political or religious values.[59] For example, evangelical Christians or conservative global warming skeptics who see Tyson's messages about evolution or climate change as conflicting with their own cultural identities may simply choose to tune him out. Both the indifference of those who see science as boring and the hostility of those who reject its conclusions may limit the impact of science documentary shows that lean toward the informational end of the spectrum—a category that includes not only cable television's *Cosmos* but also public television's *NOVA*.

Science on Public Broadcasting: The Case of *NOVA*

PBS played no role in developing Tyson's reboot of *Cosmos*—the affiliate that produced the original series sold the rights to it in 1989—but public television remains one of the leading outlets for science documentary programming to this day. In 2017, PBS was the sixth most-watched television channel overall and drew 70% more prime-time viewers than The Discovery Channel.[60] Moreover, members of the public are three times more likely to trust PBS than distrust it—making it one of the most trusted media outlets in the United States.[61]

Public television's flagship science documentary series is *NOVA*, which combines interviews with other footage to address a wide range of topics. For example, episodes from the 2017–2018 season examine the first flight around the world by a solar-powered airplane, the science of statistical forecasting, the discovery of some of the oldest human remains in the Western Hemisphere, and the impact of climate change on the Earth's weather system. Since its 1974 premiere, *NOVA* has won numerous awards.[62] It is also the most watched prime-time science series on television, with a regular audience of around 5 million viewers.[63] Our October 2016 survey found that more than a third of the US public (37%) had seen the series.

Along with *NOVA*, public television airs various other science-themed programs. In 1982, PBS launched *Nature*, an Emmy-winning wildlife series that has spanned more than three dozen seasons to date.[64] Typical episode themes include butterfly mating habits, the family life of cheetahs, and efforts to save the last male white rhino in the world. Since 2000, PBS has also aired *Secrets of the Dead*, a program that uses "the latest investigative techniques, forensic science and historical examination" to examine topic such as the Salem Witch Trials, King Tutankhamun's tomb, and the sinking of the *Titanic*.

Given that PBS is a nonprofit programming distributor, one might expect its science documentary programs to prioritize educational values over entertainment values—in contrast to the science-themed shows from privately owned, for-profit cable networks such as The Discovery Channel and Animal Planet. Yet even public television faces pressures to draw an audience. PBS itself depends on corporate sponsors, foundations, and the US government for funding, and its affiliate stations rely on donations from viewers.[65] At the same time, the makers of public television programs such as *NOVA* face pressure from another direction: their sources in the scientific community, who want these shows to help maintain a positive public image of science. After all, scientists themselves often depend on public support in conducting their research. The combined pressures on public television encourage a production approach that "blends education with both entertainment and promotion."[66]

With all this in mind, Susanna Hornig analyzed episodes of *NOVA* from 1988 to explore how the series portrays the nature of scientific work. She found that it dramatizes science as "an abstract and extremely powerful force with near-supernatural characteristics."[67] For example, an episode about superconductors describes the technology as having "almost magical power."[68] Scientists on the program appear wearing lab coats, surrounded by trappings such as equation-covered blackboards and book-filled offices that signify their "special status" as the "high priests who negotiate for us between their mysterious world and our more mundane one."[69] Hornig points out that NOVA typically portrays scientists as "*explaining* things" rather than "*doing* things"; meanwhile, nameless assistants perform any actual work shown.[70] She suggests that such portrayals may be "more likely to contribute to the mystification of science than to the demystification of it"—in contradiction to "the show's professed goal of making science accessible to nonscientists."[71]

To provide another, more recent, look at NOVA, our research assistant, D. J. McCauley, examined a dozen randomly selected episodes from 2011 to 2017. In particular, she explored whether the series portrays scientific work as dangerous along with whether it portrays scientists as working alone or in teams, as odd and peculiar, and as socially distant. Of the 12 episodes, a few highlight the potential risks of scientific inquiry. One presents researchers recreating the Montgolfier brothers' original hot air balloon ("The two men are 3,000 feet in the air, with only a fragile balloon and an open, straw-fed fire to keep them aloft."), and another depicts engineers sealing the damaged Chernobyl nuclear reactor in Ukraine ("This job, it's extremely dangerous.").[72] Most episodes, however, portray nothing more dangerous than being pinched by a fiddler crab (a "lethal weapon"—but only to other crabs) or facing mental exhaustion from competing in a robotics tournament ("It took at least a couple months for the recovery.").[73]

The NOVA episodes McCauley analyzed also emphasize the collaborative nature of scientific work. Most of them portray teams of researchers: for example, the Montgolfier brothers who built a hot air balloon, a robotics lab at the Defense Advanced Research Projects Agency (DARPA), NASA's Jet Propulsion Laboratory, and the "thirty-six scientists from 17 institutions and four countries" who excavated the Snowmass Ice Age site.[74] The scientists interviewed for the episodes typically use "we" language instead of "I" language, reinforcing the theme of science as a collective endeavor.

As for the scientists themselves, they seldom come across as either strange or distant. On the contrary, NOVA presents them as eager to share their knowledge with broader audiences. For example, one episode shows a group of scientists engaging the public through museum displays of animal remains trapped in an Ice Age lake. They even joke about the smell of a mastodon bone ("Yes, blackberry, cherry, hint of oak, hint of tobacco, yeah … clearly been in an oak barrel for three years.").[75] Other episodes depict scientists working with local carpenters to help recreate an ancient Egyptian chariot or using crowdsourcing games to solve protein-folding puzzles.[76] The narrators and interviewers usually refer to these scientists by their first names, rather than "Doctor So-and-so." Episodes also show scientists working to help society by protecting people from radioactive contamination, monitoring asteroids that could devastate the Earth, developing earthquake and tsunami warning systems, or using neuroscience to understand the minds of mass killers.[77]

While earlier research highlights the potential for *NOVA* to make science seems mysterious and inaccessible, McCauley's look at the show's more recent content suggests a different set of effects. Based on her observations, we might expect *NOVA* viewers to be particularly *unlikely* to see scientific work as dangerous, as well as particularly *unlikely* to perceive scientists as working alone or as odd and socially distant. Given how *MythBusters* and *Cosmos* emphasize the collaborative nature of scientific work over its dangers and humanize scientists as "team players" united by curiosity, we might expect similar patterns among viewers of these two programs.

A New Look at Science TV Viewing and Perceptions of Science

Our October 2016 survey of the US public provided us with data for testing these possibilities. In addition to asking respondents whether they had watched *MythBusters*, *Cosmos*, and *NOVA*, we asked them about their perceptions of science and scientists. Specifically, we asked whether they agreed or disagreed that "scientific work is dangerous," that "scientists usually work alone," that "scientists tend to look down on other people," and that "scientists tend to be odd and peculiar people."

In the case of *MythBusters*, we found no meaningful differences between viewers and non-viewers on three of the four questions (Figure 4.1). The two groups were more or less equally likely to see scientific work as dangerous (51% for non-viewers versus 49% for viewers), to see scientists as odd and peculiar (48% versus 46%), and to see scientists as looking down on other people (33% versus 31%). However, *MythBusters* viewers were 12 percentage points less likely (24%) than non-viewers to believe that scientists work alone (36%)—a gap that remained statistically significant even when we controlled other forms of media use (including *Cosmos* and *NOVA* viewing) along with key demographic factors. Although this finding doesn't prove that watching *MythBusters* caused a shift in viewers' beliefs, the relationship here fits with how the series depicts Hyneman and Savage working alongside their co-hosts and guests to conduct scientific demonstrations.

When we looked at *Cosmos* and *NOVA* viewers, we found that people who watched one tended to watch the other. Accordingly, we created a combined measure of whether respondents had seen neither, one, or both. Respondents

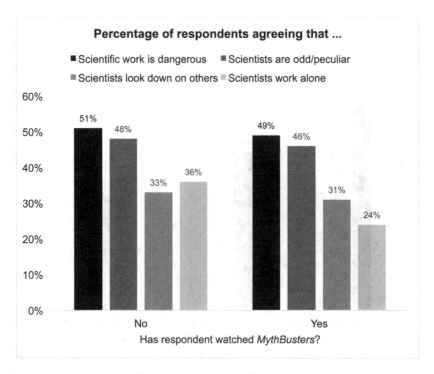

FIGURE 4.1 Perceptions of scientists and scientific work, by *MythBusters* watching (Cooperative Congressional Election Survey, 2016)

who had watched *both* programs differed from the ones who had seen *neither* on all four questions (Figure 4.2), with *Cosmos/NOVA* viewers being less likely to see scientific work as dangerous (37% versus 57%), to see scientists as odd and peculiar (32% versus 53%), to believe that scientists look down on other people (20% versus 33%), and to think that scientists work alone (13% versus 38%). All these differences were larger than 10 percentage points, and the one for scientists working alone was 25 percentage points. Furthermore, we found the same patterns after statistically controlling for other forms of media use (including *MythBusters* viewing) and key demographics. Again, our results don't conclusively prove that *Cosmos* and *NOVA* swayed audience members' perceptions of science; viewers' existing beliefs may also drive them to seek out these programs. Yet the patterns we found here suggest, at the very least, that watching the two programs goes hand in hand with an image of science as safe rather than perilous and of scientists as down-to-earth, ordinary people who work collaboratively rather than in isolation.

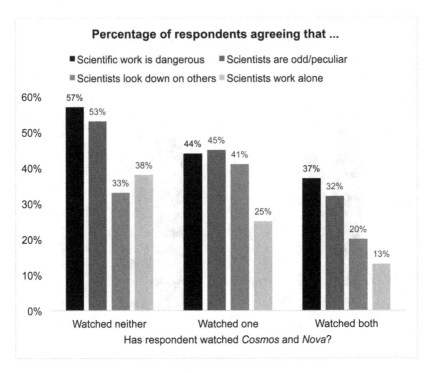

FIGURE 4.2 Perceptions of scientists and scientific work, by *Cosmos* and *NOVA* watching (Cooperative Congressional Election Survey, 2016)

This does leave aside another important issue: whether documentary television presents a diverse portrayal of the profession. We already know that movies and prime-time television tend to depict *fictional* scientists as white, heterosexual, able-bodied men.[78] So how have documentary science programs portrayed the demographics of *real* scientists? Historically, in much the same way as their fictional counterparts. For example, the NOVA episodes from 1988 that Hornig analyzed showed men, but not women, in the role of the scientist as explainer of natural mysteries.[79] Likewise, both the original *Cosmos* and the original *MythBusters* feature white men as their primary hosts, though the latter includes one woman co-host (Kari Byron) and one Asian American co-host (Grant Imahara).[80]

Some recent documentary television programs do offer counterexamples to this pattern. Most prominently, perhaps, the rebooted *Cosmos* features a Black scientist, Neil deGrasse Tyson, as the successor to Carl Sagan. Nor do the differences between the original series and the new one end there, with the latter doing more to emphasize the contributions of women scientists and scientists of color (though the majority of the scientists it discusses are white men). One episode ("Sisters of the Sun") recounts the work of Annie Jump Cannon, Henrietta Swan Leavitt, and Cecilia Payne in classifying stars, and another ("Hiding in the Light") focuses on the work of the Chinese philosopher Mozi and the Arab scientist Ibn Alhazen. Similarly, women scientists and scientists of color play the role of "explainer" in several of the *NOVA* episodes from the 2010s that McCauley analyzed. Still, such cases represent exceptions to the dominant portrayal of scientists in documentary programs.

Given that media models can shape young people's life choices, the ongoing underrepresentation of women, people of color, LGBTQ people, and disabled people in science documentaries may reinforce existing disparities in education and the workforce.[81] To be sure, audience members sometimes take inspiration from media figures across demographic divides; for example, both Tyson and NASA scientist Michelle Thaller have described Carl Sagan as a model for their own careers.[82] Yet research suggests that audience members may be particularly likely to identify with—and emulate—scientific role models who "look like them" in demographic terms.[83] If so, then a shift toward greater inclusivity within documentary television shows could help promote diversity within the broader scientific community.

Nonfiction Science Media and the Public: From Documentaries to News

Documentary science television programs are popular among the public and highly trusted by those who consume them. In some instances, the ways in which these programs combine information with entertainment can pave the way for audience members to come away with distorted perceptions—as with The Discovery Channels' *Shark Week*, which presents a skewed image of sharks as dangerous human-eaters. In other instances, infotainment programs may promote greater public understanding of scientific work—as with *MythBusters*, which emphasizes the fun and collegial nature of real-world science. Educational programs such as *Cosmos* and *NOVA* can also foster an image of scientists as sociable rather than strange and scientific work as appealing rather than scary.

Having focused thus far on television, we should mention that a wider assortment of nonfiction media play roles in bringing science to the public. For example, theatrically released documentary films have addressed topics ranging from Antarctic birds (*March of the Penguins*, 2005) to the Large Hadron Collider (*Particle Fever*, 2013). The 2006 documentary *An Inconvenient Truth*, which earned $22 million at the box office and won an Academy Award for Best Documentary, is a particularly interesting case. The film combines footage of former vice president Al Gore discussing greenhouse gas emissions with anecdotes about his life and even a clip from the animated comedy show *Futurama*. It ends with a call to action in which he describes a photograph taken by the NASA probe *Voyager 1* and invokes the original host of *Cosmos*:

> When it got 4 billion miles out in space, Carl Sagan said, "Let's take another picture of the Earth." See that pale blue dot. That's us. Everything that has ever happened in all of human history has happened on that pixel. All the triumphs and all the tragedies, all the wars, all the famines, all the major advances: it's our only home. And that is what is at stake: our ability to live on planet Earth, to have a future as a civilization.

The release of *An Inconvenient Truth* prompted considerable discussion of climate change. Within the scientific community, some experts criticized particular aspects of the movie, such as its portrayal of the certainty surrounding links between climate change and individual weather events such as Hurricane Katrina. Even so, the 19 climatologists who responded to a survey by the Associated Press all agreed that Gore "mostly got the science right."[84] Moreover, a pair of experiments conducted by Jessica Nolan suggest that people who watched *An Inconvenient Truth* came away with greater knowledge about climate change, higher levels of concern about global warming, and greater willingness to act on the issue.[85]

Nor are documentary television shows and films the only forms of nonfiction science media. Others include long-running magazines such as *Scientific American* and *National Geographic*; radio programs such as *Science Friday*; and, more recently, podcasts such as *The Infinite Monkey Cage* and *Nature Podcast*. Although these forms of media may not rival documentary television programs and films in popularity, they still reach substantial audiences. The same 2017 Pew Research Center survey we cited at the beginning of the chapter found that 25% of respondents regularly got science news from science magazines, and 12% learned about the topic from podcasts or radio programs. Furthermore, the audiences for these outlets tended to trust them: 66% said that science magazines "get the facts right when it comes to science," and 50% said so for science podcasts and radio.

Yet one type of nonfiction media overshadows all the rest when it comes to learning about science: general news outlets. A majority of the respondents in the 2017 Pew Research Center survey (54%) said they regularly got science news from "news outlets that cover a range of topics." At the same time, only 38% of the people who learned about science from general news media trusted these outlets to get the facts right about the topic.[86] Thus, general news media outlets stand out among nonfiction media for both their high popularity and low credibility as sources of scientific information. As such, this category—which includes outlets ranging from local newspapers to cable television news—merits its own in-depth look.

Notes

1 Trinidad, Elson, "September 1980: Carl Sagan's *Cosmos: A Personal Journey* airs," KCET, Sept. 10, 2014, www.kcet.org/kcet-50th-anniversary/september-1980-carl-sagans-cosmos-a-personal-journey-airs.

2 Mooney, Chris, and Sheril Kirshenbaum, *Unscientific America: How scientific illiteracy threatens our future*, Basic Books, 2009, 37.

3 *NOVA*, "The disciples of Carl Sagan," PBS, Mar. 18, 2014, www.pbs.org/wgbh/nova/blogs/secretlife/blogposts/the-disciples-of-carl-sagan/.

4 Schenider, Steve, "Cable TV notes; a channel with a difference," *New York Times*, June 16, 1985, www.nytimes.com/1985/06/16/movies/cable-tv-notes-a-channel-with-a-difference.html.

5 Chris, Cynthia, "All documentary, all the time? Discovery Communications Inc. and trends in cable television," *Television & New Media* 3, no. 1 (2002): 7–28.

6 Funk, Cary, Jeffrey Gottfied, and Amy Mitchell, "Science news and information today," Pew Research Center, Sept. 20, 2017, www.journalism.org/2017/09/20/science-news-and-information-today/.

7 Cooper, Kathryn E., and Erik C. Nisbet, "Documentary and edutainment portrayals of climate change and their societal impacts," in *Oxford Research Encyclopedia of Climate Science*, eds. Joseph E. Uscinksi, Karen Douglas, and Stephan Lewandowsky, Oxford University Press, 2017: https://oxfordre.com/climatescience/view/10.1093/acrefore/9780190228620.001.0001/acrefore-9780190228620-e-373; Evans, Suzannah, "*Shark Week* and the rise of infotainment in science documentaries," *Communication Research Reports* 32, no. 3 (2015): 265–271.

8 Lynch, Jason, "After closing scripps deal last week, Discovery sets unified upfront strategy," *AdWeek*, Mar. 12, 2018, www.adweek.com/tv-video/after-closing-scripps-deal-last-week-discovery-sets-unified-upfront-strategy/.

9 Campbell, Vincent, *Science, entertainment and television documentary*, Springer, 2016.

10 Campbell, *Science, entertainment and television documentary*; Van Dijck, Jose, "Picturizing science: The science documentary as multimedia spectacle," *International Journal of Cultural Studies* 9, no. 1 (2006): 5–24.

11 Pierson, David P., "'Hey, they're just like us!' Representations of the animal world in the Discovery Channel's nature programming," *Journal of Popular Culture* 38, no. 4 (2005): 698–712.

12 Pierson, "'Hey, they're just like us!'" 703.

13 Pierson, "'Hey, they're just like us!'" 704–705.

14 Campbell, *Science, entertainment and television documentary*.

15 Campbell, *Science, entertainment and television documentary*, 47.

16 Campbell, *Science, entertainment and television documentary*, 28.

17 Campbell, *Science, entertainment and television documentary*.

18 Wilcox, Christie, "*Venom Hunters* receives venomous backlash: Reality bites part I," *Discover*, June 21, 2016, http://blogs.discovermagazine.com/science-sushi/2016/06/21/venom-hunters-receive-venomous-backlash/.

19 Wilcox, Christie, "How committed is Discovery to no fakes?: Reality bites part IV," *Discover*, June 24, 2016, http://blogs.discovermagazine.com/science-sushi/2016/06/24/reality-bites-part-iv/#.Wstvb4jwaUk.

20 Thaler, Andrew David, "The politics of fake documentaries," *Slate*, Aug. 31, 2016, www.slate.com/articles/technology/future_tense/2016/08/the_lasting_damage_of_fake_documentaries_like_mermaids_the_body_found.html.

21 Thaler, "The politics of fake documentaries."

22 Thaler, "The politics of fake documentaries."

23 Discovery Go, "About *Shark Week*, season 2012," Oct. 24, 2020, www.discovery.com/tv-shows/shark-week/about.

24 Evans, *Shark Week*.

25 See the Appendix for details about this survey.

26 Evans, *Shark Week*.

27 Myrick, Jessica Gall, and Suzannah D. Evans, "Do PSAs take a bite out of *Shark Week*? The effects of juxtaposing environmental messages with violent images of shark attacks," *Science Communication* 36, no. 5 (2014): 544–569.

28 Otterson, Joe, "Discovery defends Michael Phelps *Shark Week* special after backlash," *Variety*, July 25, 2017, http://variety.com/2017/tv/news/michael-phelps-shark-week-discovery-1202506139/.

29 Myrick and Evans, "Do PSA take a bite out of *Shark Week*?"

30 Nosal, Andrew P., Elizabeth A. Keenan, Philip A. Hastings, and Ayelet Gneezy, "The effect of background music in shark documentaries on viewers' perceptions of sharks," *PLOS One* 11, no. 8 (2016): e0159279.

31 Nosal et al., "The effect of background music," 13.

32 Myrick and Evans, "Do PSA take a bite."

33 Thaler, Andrew David, and David Shiffman, "Fish tales: Combating fake science in popular media," *Ocean & Coastal Management* 115 (2015): 88–91.

34 Jenkins, Henry, "Transmedia storytelling and entertainment: An annotated syllabus," *Continuum* 24, no. 6 (2010): 943–958.

35 Discovery Go, "About MythBusters," Oct. 28, 2020, https://go.discovery.com/tv-shows/mythbusters/about.

36 Collins, Scott, "Obama episode of *MythBusters* draws 2.2 million viewers," *Los Angeles Times*, Dec. 9, 2010, http://latimesblogs.latimes.com/showtracker/2010/12/obama-episode-of-mythbusters-draws-22-million-viewers.html.

37 Television Academy, "*MythBusters*," Oct. 28, 2020, www.emmys.com/shows/mythbusters.

38 Schwartz, John, "The best science show on television?" *New York Times*, Nov. 21, 2006, www.nytimes.com/2006/11/21/science/21myth.html.

39 Lecher, Colin, "Watch Obama say goodbye to the *MythBusters*," *The Verge*, Mar. 7, 2016, www.theverge.com/2016/3/7/11171982/mythbusters-series-finale-president-barack-obama-goodbye.

40 Zavrel, Erik A., "How the Discovery Channel television show *MythBusters* accurately depicts science and engineering culture," *Journal of Science Education and Technology* 20, no. 2 (2011): 201–207; Zavrel, Erik, "Pedagogical techniques employed by the television show *MythBusters*," *The Physics Teacher* 54, no. 8 (2016): 476–479.

41 Zavrel, Erik, and Eric Sharpsteen, "How the television show *MythBusters* communicates the scientific method," *The Physics Teacher* 54, no. 4 (2016): 228–232.

42 Dunning, Brian, "*MythBusters*: Where is the MythBusting?" SkepticBlog, Sept. 2, 2010, www.skepticblog.org/2010/09/02/mythbusters-mythbusting/.

43 Kirby, David A., "The changing popular images of science," in *The Oxford handbook of the science of science communication*, ed. Kathleen Hall Jamieson, Dan Kahan, and Dietram A. Scheufele, Oxford University Press, 2017: 295–296.

44 Kirby, "The changing popular images of science," 295–296.

45 Zavrel, "How the Discovery Channel."

46 Hess, David J., *Alternative pathways in science and industry: Activism, innovation, and the environment in an era of globalizaztion*, MIT Press, 2007.

47 Brossard, Dominique, and Bruce V. Lewenstein, "A critical appraisal of models of public understanding of science," in *Communicating science: New agendas in communication*, eds. LeeAnn Kahlor and Patricia Stout, Routledge, 2010: 11–39; Irwin, Alan, and Brian Wynn, eds., *Misunderstanding science? The public reconstruction of science and technology*, Cambridge University Press, 2003; Ley, Barbara L., *From pink to green: Disease prevention and the environmental breast cancer movement*, Rutgers University Press, 2009.

48 During the COVID-19 pandemic, Savage cited one of the show's demonstrations to encourage mask-wearing. Adam Savage, May 13, 2020, https://twitter.com/donttrythis/status/1260748699728089088?lang=en; Zavrel, "How the Discovery Channel television show *Mythbusters*."

49 Campbell, *Science, entertainment and television documentary*.

50 Davis, Nicola, "*Cosmos*: How the creator of *Family Guy* remade Carl Sagan's pivotal TV series," *Guardian*, Apr. 8, 2014, www.theguardian.com/technology/2014/apr/08/cosmos-carl-sagan-seth-macfarlane-family-guy.

51 Davis, "*Cosmos*: How the creator of *Family Guy*."

52 Kissell, Rich, "*Cosmos* draws biggest global audience ever for National Geographic Channel," *Variety*, July 7, 2014, http://variety.com/2014/tv/news/cosmos-draws-biggest-global-audience-ever-for-national-geographic-channel-1201257111/.

53 Television Academy, "*Cosmos: A SpaceTime Odyssey*," Oct. 28, 2020, www.emmys.com/shows/cosmos-spacetime-odyssey.

54 The new episodes were originally slated to premiere in March 2019, but the networks delayed the program's return in response to allegations of sexual misconduct on Tyson's part. On March 15, 2019, Fox and National Geographic announced that they had completed their own investigation and were planning to move forward with the program.

55 Cruz, Giblert, "Q&A: Astrophysicist Neil deGrasse Tyson," *TIME*, Jan. 21, 2009, http://content.time.com/time/health/article/0,8599,1872621,00.html.

56 Of the respondents, 38% viewed him favorably while 23% viewed him unfavorably and another 38% hadn't heard of him or couldn't rate him.

57 Akin, Heather, Bruce Hardy, Dominique Brossard, Dietram A. Scheufele, Michael A. Xenos, and Elizabeth A. Corley, "The pitfalls of popularizing science beyond the proverbial choir: Lessons from *Cosmos* 2.0," Paper presented at the annual meeting of the American Association for the Advancement of Science, Washington, DC. 2016.

58 Stiffman, Eden, "KQED to study how to reach and engage millennials with science media," *Current: News for People in Public Media*, Apr. 29, 2019, https://current.org/2019/04/kqed-to-study-how-to-reach-and-engage-millennials-with-science-media/.

59 Kahan, Dan M., "Evidence-based science filmmaking initiative: Study no. 1," Jan. 11, 2016, https://ssrn.com/abstract=2713563.

60 PBS, "About PBS," Apr. 26, 2018, www.pbs.org/about/about-pbs/overview/.

61 Mitchell, Amy, Jeffrey Gottfried, Jocelyn Kiley, and Katrina Eva, "Political polarization and media habits," Pew Research Center, Oct. 21, 2014, www.journalism.org/2014/10/21/political-polarization-media-habits/.

62 PBS, "Awards," Oct. 28, 2020, www.pbs.org/wgbh/nova/about/tvaw.html.

63 PBS, "About *NOVA*," Oct. 28, 2020, www.pbs.org/wgbh/nova/about/.

64 PBS, "*Nature*: About the series," Oct. 28, 2020, www.pbs.org/wnet/nature/about/.

65 Hornig, Susanna, "Television's *NOVA* and the construction of scientific truth," *Critical Studies in Media Communication* 7, no. 1 (1990): 11–23.

66 Hornig, "Television's *NOVA*," 14.

67 Hornig, "Television's *NOVA*," 21.

68 Hornig, "Television's *NOVA*," 15.

69 Hornig, "Television's *NOVA*," 16–18.

70 Hornig, "Television's *NOVA*," 17, 22.

71 Hornig, "Television's *NOVA*," 21–22.

72 *NOVA*, "Ben Franklin's Balloons" (Season 41, Episode 17); "Building Chernobyl's Mega-Tomb" (Season 44, Episode 8).

73 *NOVA*, "Extreme Animal Weapons" (Season 44, Episode 26); "Rise of the Robots" (Season 43, Episode 8).

74 *NOVA*, "Ben Franklin's Balloons"; "Rise of the Robots"; "Asteroid: Doomsday or Payday?" (Season 40, Episode 24); "Ice Age Death Trap" (Season 39, Episode 5).

75 *NOVA*, "Ice Age Death Trap."

76 *NOVA*, "Building Pharaoh's Chariot" (Season 40, Episode 5); "The Origami Revolution" (Season 44, Episode 5).

77 *NOVA*, "Building Chernobyl's Mega-Tomb"; "Asteroid: Doomsday or Payday?"; "Japan's Killer Quake" (Season 38, Episode 9); "The Mind of a Killer" (Season 40, Episode 7).

78 See Chapters 2 and 3.

79 Hornig, "Television's *NOVA*."

80 Zavrel, "How the Discovery Channel."

81 O'Keeffe, Moira, "Lieutenant Uhura and the drench hypothesis: Diversity and the representation of STEM careers," *International Journal of Gender, Science and Technology* 5, no. 1 (2013): 4–24; Steinke, Jocelyn, "Cultural representations of gender and science: Portrayals of female scientists and engineers in popular films," *Science Communication* 27, no. 1 (2005): 27–63.

82 PBS, "The Disciples of Carl Sagan," Oct. 28, 2020, www.pbs.org/wgbh/nova/blogs/secretlife/blogposts/the-disciples-of-carl-sagan/.

83 O'Keeffe, "Lieutenant Uhura and the drench hypothesis"; Steinke, "Cultural representations of gender and science."

84 Borenstein, Seth, "Scientists OK Gore's movie for accuracy," *Washington Post*, June 27, 2006, www.washingtonpost.com/wp-dyn/content/article/2006/06/27/AR2006062700780.html.

85 Nolan, Jessica M., "*An Inconvenient Truth* increases knowledge, concern, and willingness to reduce greenhouse gases," *Environment and Behavior* 42, no. 5 (2010): 643–658.

86 See also Brewer, Paul R., and Barbara L. Ley, "Whose science do you believe? Explaining trust in sources of scientific information about the environment," *Science Communication* 35, no. 1 (2013): 115–137.

5

SCIENCE NEWS

Tucker Carlson: I think most people are open to the idea that climate is changing. It has always changed, by the way ... the core question, as far as I can tell, is why the change. Is it part of the endless cycle of climate change, or is human activity causing it? That seem to be the debate to me, and it seems an open question, not a settled question ...

Bill Nye: It's not an open question. It's a settled question. Human activity is causing climate change.

Carlson: To what degree?

Nye: To a degree that it's a very serious problem in the next few decades ...

Carlson: So much of this you don't know. You pretend that you know, but you don't know.

—*Tucker Carlson Tonight*, Fox News Channel (February 27, 2017)

Chris Cuomo: NASA scientists say 97% of climate scientists agree that climate warming trends are extremely likely due to human activities.

—*Cuomo Prime Time*, CNN (October 17, 2018)

Two media worlds collided on August 3, 2011: cable news and children's television. In a *Fox and Friends* segment, host Gretchen Carlson told viewers that the Department of Education was using SpongeBob SquarePants, a character from a popular program on the Nickelodeon network, to teach students about global warming. "The government agency showed kids this cartoon ... that blamed man for global warming," she explained, "but they did not tell kids that that is actually a disputed fact. Oops!" For his part, co-host Steve Doocy argued

DOI: 10.4324/9781003190721-5

that "clearly Nickelodeon is pushing a global warming agenda." He went on to say that while there was "no disputing the fact that the Earth is getting a little warmer, the big question is, is it man-made or is it just one of those gigantic, climactic phases we're going through... There's science on both sides, there are a lot of scientists who say it's this, others say it's that."

Six days later, another cable television outlet entered the fray: *The Colbert Report with Stephen Colbert*, a late-night satirical news program on the Comedy Central network. In his ironic persona of a conservative talk show host, Colbert declared that he "refuse[d] to buy into the myth of global warming. It's just another big media lie...And these hot air-heads have started indoctrinating our kids." After running a clip of Carlson's comments about SpongeBob, he added, "This is how it works, folks. The liberal media brainwashes our kids by sneaking propaganda into their cartoons in a code only kids can understand ... obviously, the real issue here is the way SpongeBob is treating as accepted fact the *theory* that humans caused global warming." Colbert concluded by echoing—and, in the process, mocking—Doocy's analysis. "Yeah, make up your mind, scientists," he scoffed. "Do you believe in the conclusions you've reached from decades of peer-reviewed study?"

Taken together, these two television segments illustrate how different media outlets can interpret—or *frame*—science in dramatically different ways. Whereas *Fox and Friends* suggested that scientists were divided on the causes of global warming, *The Colbert Report* used satire to highlight the scientific consensus on climate change and poke fun at the Fox hosts' charges of media bias. Such contrasts extend well beyond the mini-controversy over SpongeBob. For example, Fox News Channel follows a pattern of emphasizing skepticism about global warming, while the other two major cable news networks—CNN and MSNBC—tend to present human-caused climate change as a scientific certainty. These clashing messages, in turn, could influence whether audience members accept or doubt the existence of global warming and, ultimately, whether they support action to address it.

The same dynamic may play out for other science-related subjects, including ones where—as with climate change—wide gaps separate what scientists say and what many members of the public think: for example, whether humans evolved from other forms of life, whether genetically modified foods are safe to eat, and whether vaccines are safe for children.[1] For each of these issues, news media messages could serve to widen or bridge the divides between scientists and non-scientists—and, in the process, shape what choices we, as a society, make about policy issues such as teaching evolution in schools, labeling genetically modified organisms (GMOs) in foods, and requiring vaccinations.

With this in mind, we look at how news outlets have covered a range of topics involving science, from climate change to coronavirus disease 2019 (COVID-19). We also explore whether Americans' perceptions of four prominent topics—global warming, evolution, GMO foods, and vaccines—reflect where they get information about science, along with how they make sense of such information based on their own worldviews and experiences. First, though, we should say more about how the news media go about covering science and why their messages may influence audience members. To do that, we build on the foundation of framing theory.

Framing Science

The universe is complicated. So is the Earth, and so are humans themselves. As a result, any field of scientific inquiry inevitably uses theories, methods, and even language unfamiliar to the average member of the public—or, for that matter, the average journalist. For example, astronomers talk about dark energy, while physicists discuss Higgs bosons. This complexity creates a challenge for media practitioners, who need to find ways to tell stories about science that their audience members will understand.

One solution to the problem lies in framing, a concept with roots that span the social sciences. Working from one direction, sociologists such as Erving Goffman have argued that people use interpretive frameworks to make sense of the world around them.[2] Working from another angle, psychologists such as Daniel Kahneman and Amos Tversky have found that subtle variations in the phrasing of a choice can influence people's judgments about it.[3] Political scientists and communication scholars, in turn, have combined both lines of research to show that the ways in which news outlets frame stories can shape public opinion about a wide range of topics—including scientific ones.

So, what exactly are frames? To echo William Gamson and Andre Modigliani, they're storylines for what an issue is "all about," built from catchphrases, metaphors, images, and symbols.[4] Scientists, activists, politicians, interest groups, journalists, and even ordinary citizens can frame virtually any given issue in multiple ways, some of which may be diametrically opposed to one another. Take nuclear power, which could be framed as an alternative to climate-damaging fossil fuels or as a dangerous technology with the potential for runaway disasters.

As Robert Entman points out, these frames can exist on four different levels: in the minds of messengers, such as reporters and talk show hosts; in media messages, from print stories to cable news segments; in the minds of receivers, be they *New York Times* readers or Fox News viewers; and in the broader culture that provides us with a set of shared frames.[5] When journalists and other media practitioners frame a given issue by emphasizing a specific understanding of what caused it, what's at stake in it, or what to do about it, audience members may accept these interpretations and use them to form their own judgments.

Researchers have looked at many different aspects of how the news media frame issues, including topics related to science. One key choice journalists and other communicators make in telling stories is whether to focus on the big picture (thereby using a *thematic* frame) or an individual case (thereby using an *episodic* frame).[6] For example, a news story about how climate change has affected polar bears could focus on trends in bear populations or on the tale of one bear struggling to survive in a changing environment.[7] Likewise, a news story about shark attacks against humans could focus on the overall rate of such attacks or on a single vivid instance of a shark killing a swimmer.[8]

In framing an issue, journalists and other media practitioners also make decisions about what aspects of it to emphasize. Looking across a range of different topics, Matthew Nisbet has identified a set of common frames in science coverage.[9] This typology includes the progress frame, which revolves around the potential for new scientific and technological developments to solve problems and improve the quality of life, as well as the runaway science frame, which raises the specter of an out-of-control "Frankenstein's monster" or a "Pandora's box" full of unexpected problems. Other common science frames are the morality or ethics frame (what's right or wrong); the uncertainty frame (what's known or unknown); and the strategy or game frame (who's winning, who's losing, and what tactics they're using in a political or legal battle).

The case of stem cell research illustrates how news outlets can use a range of frames in covering scientific issues. The successful isolation of human embryonic stem cells in 1998 sparked a wave of studies on their potential applications—along with a surge of media stories about the topic. Some of these stories focused on potential medical benefits from stem cell research (a progress frame), while others presented dire images of mad scientists or Nazi doctors engaging in medical abuses (an ethics frame).[10] Coverage also framed the issue in terms of the unknown outcomes of stem cell research (an uncertainty frame) or the political fight between Republicans and Democrats over funding and regulations for such research (a strategy frame). Each of these frames offers a different way of interpreting embryonic stem cell research—and of judging whether the government should support or ban it.

More recently, news coverage of COVID-19 has featured its own diverse set of frames, including ones highlighting economic consequences, government responses, and political impacts.[11] As in the case of stem cell research, these frames suggest alternative ways to understand and form opinions about the pandemic. For example, economic frames may lead audience members to judge public health responses in terms of potential business closures and job losses, while public accountability frames invite audience members to judge government officials for their successes or failures in handling the pandemic.

What Shapes News Framing?

Faced with so many possible frames, media storytellers need ways to choose among them. One professional standard for guiding such decisions is the norm of objectivity, which calls for journalists to cover issues in an unbiased manner.[12]

In practice, reporters often try to adhere to this principle through coverage that presents "both sides" of an issue and lets audience members draw their own conclusions about who's right.[13] For example, news stories about the potential discovery of "cold fusion" in 1989 often featured both supportive and critical perspectives—at least until subsequent research failed to replicate the initial findings.[14] Similarly, our content analysis of news about the chemical bisphenol A (BPA) from 1996 to 2009 revealed that stories often gave equal weight to two possibilities: that it's safe to use in plastic products, and that it poses risks to human health.[15] On each issue, coverage seemingly upheld the norm of objectivity by offering a "balanced" take.

Yet a "he said, she said" style of reporting may lead news outlets to frame the science around an issue as uncertain even when the evidence strongly supports one conclusion over another. Consider the initial news media response to a 2002 cloning hoax involving a UFO cult known as the Raelians. In covering this scientific fraud, some journalists—including CNN's Sanjay Gupta—repeated claims from Clonaid, the sham company purporting to have produced a human baby through cloning, alongside commentary from actual biomedical scientists.[16] Several years later, a flurry of news stories about the Large Hadron Collider followed a similar pattern by pairing the views of mainstream physicists with unfounded speculations about how the particle accelerator might create a world-destroying black-hole (it didn't) or produce a time loop sabotaging the collider itself (that didn't happen, either).[17]

Economic pressures can also shape how news organizations cover scientific issues. Commercial outlets such as newspapers and cable networks depend on subscriptions and advertising revenue to survive, while even public media outlets such as PBS and NPR affiliates rely on donations. The need to draw an audience may push these organizations to cover scientific topics in ways that highlight novelty, drama, and conflict—all of which make for compelling stories. For example, coverage of artificial intelligence has sensationalized the (unlikely) possibility that the technology will lead to a "revolt of the machines" along the lines of the ones in films such as *2001: A Space Odyssey*, *Terminator*, and *The Matrix*.[18] Likewise, news stories about nanotechnology have created drama by focusing on the (improbable) "gray goo" scenario in which self-replicating nanobots destroy all life on earth.[19]

Last, but not least, news sources can shape which frames appear in science coverage. Along with scientists themselves, a host of interest groups, politicians, corporations, and activists may seek to promote their own preferred frames through the media. Take global warming, where fossil fuel industry spokespeople and even US presidents have pushed skeptical frames.[20] Or evolution, where advocates of creationism and "intelligent design" (ID) have challenged scientists to debate the topic.[21] Or GMOs, where activist groups such as Greenpeace have popularized the "Frankenfood" label in news coverage.[22] Or vaccines, where celebrities such as Jenny McCarthy have won a media spotlight for spurious claims of links to autism.[23]

Understandably, many in the scientific community see the competition to frame issues in the media as a troubling process.[24] Some may argue that science communicators should simply "rely on the data" instead of using "spin" to frame issues, but news outlets can't "just stick to the facts."[25] Journalists tell *stories*, and frames give them the means to do so. Viewed in this light, framing isn't a distortion of the communication process; it's a fundamental part of it. Though specific frames may promote misinterpretations, framing itself is what provides meaning to science coverage—or, for that matter, scientists' own messages.

Framing Effects

Just as frames give journalists and other media practitioners tools for explaining complex scientific topics, they give audience members ways to understand these topics. Indeed, several decades' worth of studies have found that news framing can influence audience members' perceptions of science—at least some of the time.

Much of the evidence for framing effects comes from experiments. For example, Sol Hart found that a news story with a big-picture frame ("Thousands of polar bears struggle for food in the Arctic") did more than an individualized story ("Polar bear struggles to survive in the Arctic") to build support for government action on climate change.[26] At first glance, it may seem surprising that a vivid tale about one polar bear did less to move audience members. Yet the big picture story encouraged them to see the issue of climate change as the *government's* responsibility—and, in doing so, drove support for policy solutions.

Survey-based studies have also captured links between how the news media frame scientific issues and what audience members think about these issues. Take stem cell research and nanotechnology, where the lion's share of coverage has emphasized "progress" frames about benefits to society over "mad science" or "Pandora's box" frames. On the first issue, Nisbet and his colleagues found that respondents who paid attention to science news were especially supportive of stem cell research.[27] Likewise, Dietram Scheufele and Bruce Lewenstein found that people who followed science news had particularly favorable views of nanotechnology.[28]

Framing effects can even extend beyond what people think to how they talk to one another about issues. When William Gamson conducted a series of focus group discussions among ordinary citizens, he observed them invoking media frames for nuclear power. For example, his participants used "runaway science" frames when discussing real-life accidents such as Chernobyl along with fictional examples from movies such as *Silkwood* and *The China Syndrome*.[29] Put simply, news framing can set the terms of conversations among citizens.[30]

The logic of framing suggests that whoever succeeds at defining an issue in news coverage has an advantage in shaping what society does about that issue. Even so, frames don't automatically change minds. When audience members come to news stories with their own beliefs in mind, they may reject media frames that don't fit those beliefs.[31] Furthermore, competing frames within news coverage can "cancel out" the effects of one another.[32] Thus, the frames most likely to sway public opinion are the ones that dominate the coverage audience members follow *and* resonate with their existing beliefs.

The Audience for Science News

To set the stage for looking at whether—and, if so when—public perceptions reflect frames in news coverage of science-related issues, we surveyed a nationally representative sample of Americans in November 2016 and asked them where they got information about science. Around a third of our respondents said they learned about it from newspapers (33%). Similar percentages relied on science magazines such as *Scientific American* and *National Geographic* (30%) and on the evening news programs of the broadcast networks ABC, CBS, and NBC (32%). Respondents also got information about science from CNN (19%), Fox News (16%), and MSNBC (12%). In short, members of the US public learn about science from a variety of sources, with no single outlet dominating. Meanwhile, a sizable "missing audience" for science information doesn't learn about the topic from any news sources and, as a result, doesn't directly receive any news frames for science issues.[33]

Beneath these patterns, Americans' news habits are polarized along partisan lines in ways that suggest selective exposure—that is, choosing outlets that reinforce one's own point of view and avoiding outlets that don't.[34] Our poll revealed clear differences in where respondents learned about science depending on whether they identified as Democrats, Republicans, or independents. Democrats were particularly likely to get information about science from newspapers (39%), broadcast network evening news (42%), CNN (28%), and MSNBC (20%) whereas Republicans were especially likely to learn from Fox News (33%). Still, the consumption of science news isn't *completely* polarized along party lines. For example, one out of every ten Democrats got science information from Fox News (10%), and around the same percentage of Republicans did so for CNN (11%).

In exploring the links between these media habits and public beliefs about science, we focus on the topics of climate change, evolution, GMOs, and vaccines. Each of these issues has received considerable media attention over the past two decades, and each is one where many members of the public reject dominant scientific views. At the same time, our cases differ from one another not only in terms of the patterns of coverage they've received but also the degree to which they map onto political and religious fault lines. Thus, a look at all four issues should give us a better sense of whether—and, if so, when and how—Americans' perceptions of science-related issues reflect the news messages they consume, along with their preexisting beliefs.

News about Climate Change

Climate change may well be the most prominent, and contentious, science-related issue of the past 30 years. Over the course of the 1980s and 1990s, attention to the topic expanded from scientific circles to the political arena, driven by developments such as the 1997 negotiations over the Kyoto Protocol (a treaty to reduce greenhouse gases).[35] In response, US news outlets began devoting more coverage to the topic while also shifting from a focus on scientists' frames to include more frames from politicians and interest groups.[36]

The 2001 report from the Intergovernmental Panel on Climate Change helped trigger a new wave of media attention, as the number of television news and newspaper stories about global warming spiked in the months immediately following the report's release.[37] Coverage declined over the next few years, only to surge again in 2006–2007 and reach new heights by the end of the decade. After 2010, news attention to climate change rose and fell with events, but at a substantially higher level on average than in the 2000–2005 period.

The content of this coverage has reflected ongoing efforts by different groups to frame the issue.[38] For example, fossil fuel industry allies and Republican leaders have promoted two frames designed to undermine support for government action on climate change: an economic consequence frame highlighting potential financial hardships from policies to curb greenhouse gas emissions, and an uncertainty frame presenting the science on the issue as "unsettled." Meanwhile, environmental groups and Democratic politicians have promoted a Pandora's box frame emphasizing the urgency of the dangers posed by global climate change, from rising sea levels to extreme weather, and a public accountability frame casting their opponents as waging a "war on science." Such competing public relations efforts have also provided journalists with fodder for one of their own favored storylines: strategy framing focused on winners, losers, and tactics in the political battle over the topic.

At first, the push to frame climate change as uncertain seemed to sway the tenor of media coverage. Whereas surveys of scientists and reviews of climate change studies have consistently shown an overwhelming consensus that global temperatures are rising due to human activity, early US news stories about the topic portrayed a more evenly divided debate.[39] In a pair of content analyses, Maxwell Boykoff and his collaborators found that newspaper articles from 1988 to 2002 and television news segments from 1996 to 2004 tended to balance claims that humans are causing global warming with claims that any change in the climate reflects "natural fluctuations."[40] Boykoff concluded that journalists' attempts to tell "both sides" of the issue yielded a distorted picture of the scientific evidence.[41]

Yet news framing of scientific issues can change over time. In the case of climate change, Boykoff found that newspaper coverage from 2005 and 2006 shifted toward emphasizing the scientific consensus.[42] This trend may have reflected media discussions of events such as the devastation from Hurricane Katrina in 2005 and the release of the documentary *An Inconvenient Truth* the following year. Similarly, Matthew Nisbet found that newspaper coverage of climate change tended to use "consensus framing" in 2009 and 2010—a period that encompassed a United Nations meeting on climate change, a controversy over leaked emails from climate scientists, and an unsuccessful push by President Barack Obama to implement a "cap-and-trade" program limiting greenhouse emissions.[43]

News coverage of climate change has differed substantially across outlets, as well. In particular, Lauren Feldman and her colleagues found that Fox News coverage from 2007 and 2008 was far more likely than CNN and MSNBC coverage from the same period to take a dismissive tone toward climate change and feature global warming skeptics as interview guests.[44] A study of coverage from 2011 also showed that Fox News tended to dismiss concerns over climate change as "political correctness" whereas the broadcast networks, CNN, and MSNBC devoted more coverage to the causes and effects of global warming.[45]

Such contrasting messages in news coverage can shape people's thoughts, feelings, and intentions about global warming. For example, Teresa Myers and her colleagues found that audience members reacted differently depending on whether they saw news segments framing climate change in terms of public health or national security.[46] Likewise, Matthew Feinberg and Rob Willer showed that framing climate change as a moral issue reduced the gap between how conservative and liberal audience members viewed the topic.[47]

Still, not all frames sway all audience members. When Julia Corbett and Jessica Durfee gave participants in their study a news story that framed climate change as controversial, it generated less uncertainty among those who already held pro-environmental beliefs.[48] Erik Nisbet and Sol Hart even found that news coverage highlighting the negative effects of global warming produced "boomerang effects" among Republicans, pushing them to *oppose* action on the issue.[49] In this case, audience members with strong prior beliefs may have responded by mentally reinforcing their existing views—a process psychologists call *motivated reasoning*.

More broadly, Americans' perceptions of climate change reflect their media habits in ways that dovetail with how news outlets have framed the issue. When

Hart and his colleagues analyzed survey data from 2009, they found that the people who paid the most attention to science news were the ones most likely to see climate change as harmful.[50] Beneath such overall patterns, however, different outlets can produce different effects. Looking at survey data from 2008, Feldman and her collaborators showed that Fox News viewers were especially likely to reject the scientific consensus on global warming; meanwhile, CNN and MSNBC viewers were particularly likely to accept it.[51] She and her team also found that conservative media use and global warming skepticism reinforced one another in a feedback loop.[52]

Our 2016 survey allowed us to take a new look at whether Americans' perceptions of climate change reflect their news media habits. In addition to asking respondents where they learned about science, we asked them whether they believed that "the earth is getting warmer because of human activity." Most forms of science news consumption went hand in hand with belief in human-caused climate change (Figure 5.1). For example, we found a 20-point gap between CNN viewers (86% of whom believed in human-caused climate change) and

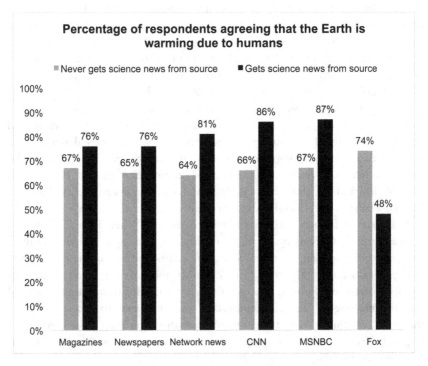

FIGURE 5.1 Beliefs about global warming, by news media use (Cooperative Congressional Election Survey, 2016)

non-viewers (66%), as well as an equally wide gap between MSNBC viewers (87%) and non-viewers (67%). Similar patterns emerged for science magazines, newspapers, and broadcast network news. Fox News was the one exception: only half of its viewers (48%) believed in human-caused climate change, compared to three-fourths (74%) of non-viewers.

Given that people's political and religious views can shape both their media habits *and* their perceptions of climate change, the patterns we found could reflect the influence of news framing, the polarization of media audiences, or both. Accordingly, we also tested how each form of media use was related to beliefs about climate change after statistically controlling for the other forms of media use along with partisanship, ideology, religiosity, and demographics. This time, we found three clear patterns. First, learning about science from CNN or MSNBC was linked to believing in human-caused climate change. Second, so was getting news about science from newspapers, reflecting the trend toward consensus coverage in print news. Third, learning about science from Fox News was linked to disbelieving in human-caused climate change.

In short, ties between media habits and climate perceptions have carried over into the 2010s. Viewers of Fox News, which has led the way in dismissing climate change, still stand out from the rest of the public for their rejection of the scientific consensus on human-caused global warming. Meanwhile, the audiences for newspapers and the other two cable news networks are especially likely to accept this consensus.

Coverage of the 2018 National Climate Assessment illustrates the continuing tensions in media framing of climate change. When the US government issued this report, news attention to global warming spiked. Much of the resulting coverage focused on the conclusions reached by scientists from 13 different federal agencies: namely, that the earth's climate is changing due to human activity and that climate change will cost many lives, as well as hundreds of billions of dollars, unless swift and extensive action is taken.[53] However, television news shows also featured opposing frames from non-scientists such as Republican politician Rick Santorum, who claimed that climate scientists are "driven by the money that they receive."[54]

News about Evolution

News coverage of evolution and its doubters goes back much further than news coverage of climate change. The scientific community has long accepted the conclusion—proposed by Charles Darwin in his 1859 book *On the Origin of Species*—that all forms of life, including humans, developed through natural selection. To this day, however, many Americans reject the scientific consensus on evolution in favor of religious-based explanations for the origins of life.[55]

One such account is "young Earth creationism," which draws on a literalist interpretation of the Bible to assert that the planet is only a few thousand years old and that humans were created in their present form by God.

Undoubtedly the most dramatic media event involving evolution was the 1925 trial of John Thomas Scopes for breaking a Tennessee law against teaching evolution in the public schools. The "Scopes Monkey Trial," as it came to be known, developed (or, perhaps, devolved) into a media circus when famed attorney Clarence Darrow agreed to speak for the defendant and former presidential candidate William Jennings Bryan joined the prosecution. Newspapers, magazines, and radio stations gave the case extensive coverage, much of it framing the proceedings in terms of strategy and hoopla rather than science (primate-related humor was another common theme).[56] The prosecution prevailed in the trial itself—at least until a higher court overturned the verdict on a technicality—but the court of media coverage was often more sympathetic to the defense.

Prohibitions against teaching evolution remained on the books in some states until a 1968 US Supreme Court ruling. Afterward, anti-evolution forces adapted to their changing environment with a new strategy: pushing for "scientific creationism" to be taught alongside evolution. They appealed to the journalistic norm of objectivity by calling for a "balanced" presentation of their claims alongside "Darwinism"—and, in doing so, gained exposure for their views in the national news media.[57] By the early 1980s, anti-evolution activists had successfully lobbied several states to pass laws requiring equal time in the public schools for a Biblical-based account of life's origins, now cloaked in the trappings of science. These victories didn't last long, however. Laws promoting "creation science" went extinct in 1987, when a US Supreme Court decision struck them down as unconstitutional attempts to establish a public religion.

Stymied again, anti-evolution forces came up with yet another approach: promoting ID. Unlike scientific creationism, ID—as presented in textbooks such as *Of Pandas and People*—omits any direct references to God or the Bible; instead, it revolves around claims that the complexity of life implies some sort of intelligent architect. A campaign led by the Discovery Institute, a fundamentalist Christian think tank, succeeded in swaying the Ohio and Kansas school boards to adopt standards that opened public classrooms to the anti-evolution arguments of ID.[58] Furthermore, the local school board of Dover, Pennsylvania voted in 2004 to require that students be read a statement about ID. When a group of Dover parents sued the board, the case drew national media attention. In this round of the "evolution wars," scientists and their allies framed ID as a "Trojan horse" for sneaking creationism into public schools.[59] For their part, the Discovery Institute framed evolution and ID as two equally legitimate sides of a scientific debate by arguing that schools should "teach the controversy."[60]

Playing to the norm of objectivity seemed to pay dividends for ID proponents: many news stories about the Dover, Ohio, and Kansas school boards'
actions balanced pro-evolution frames with pro-ID frames. In one study, Jason
Rosenhouse and Glenn Branch found that cable news outlets often treated evolution and ID as two competing scientific viewpoints—and that Fox News went
even further, with programs such as *The O'Reilly Factor* framing evolution as
"inherently atheistic."[61] In another study, Chris Mooney and Matt Nisbet found
that opinion pieces in local Dover newspapers were almost evenly split between
support for teaching ID and support for teaching only evolution.[62] Looking at
a dozen newspapers in Kansas, Pennsylvania, and Ohio, Joshua Grimm saw the
same pattern of stories balancing pro-ID and anti-ID frames.[63]

As in previous flare-ups over the teaching of evolution, the federal courts
ultimately ruled in favor of its defenders. The 2005 decision in *Kitzmiller v. Dover
Area School District* called ID "a mere re-labeling of creationism ... not a scientific
theory" and struck down the school board's policy. In the wake of the ruling, Ohio and Kansas abandoned their previously adopted standards, and media
attention to the topic faded. Still, the evolution wars didn't end there. Since
2005, anti-evolution activists have continued to push for legislation promoting
ID under the mantles of "academic freedom" and "critical thinking."[64] In 2008,
Louisiana passed a "Science Education Act" allowing anti-evolution perspectives into the classroom, and Tennessee followed suit four years later. Both laws
remained on the books as of 2020.

Over the past decade, evangelist and young Earth creationist Ken Ham has
emerged as another prominent critic of evolution. His organization, Answers
in Genesis, launched a "Creation Museum" in 2007 and an "Ark Encounter"
(based on Noah's Ark, from the Biblical account of the flood) in 2016. He
also made news in 2014 when science communicator and television show host
Bill Nye "the Science Guy" agreed to debate him at the Creation Museum.
Some in the scientific community voiced concerns that the event would
merely help Ham publicize his views, but Nye saw it as a chance to expose
creationism as "bad for science education, bad for the U.S., and thereby bad for
humankind."[65]

The debate took place on February 4, 2014, and received considerable
media attention. News coverage largely framed it in strategy terms—who won,
whether Ham enjoyed a "home team" advantage, and so on—while presenting
both Nye's and Ham's arguments. For example, ABC's *World News Tonight* talked
about "game time" and concluded that "the debate rages on." By presenting Nye
and Ham as two combatants in an unsettled controversy instead of highlighting
the scientific consensus on the topic, such coverage may have reinforced public
perceptions that scientists are divided on evolution.

This raises a question: can anti-evolution forces use media framing to shape
how audience members think about evolution and its place in science education?

Maybe so, judging by an experiment conducted by Thomas Nelson and his colleagues. They found that framing ID in terms of fairness—as the Discovery Institute did in its efforts to sway news coverage—boosted support for teaching ID alongside evolution.[66] Specifically, this frame influenced audience members by shaping how much importance they attached to the principle of fairness in judging arguments about ID.

At the same time, scientists and their allies can counter such framing effects when given the chance to explain how scientific reasoning works. In a follow-up study, Nelson and his team found that college students were less likely to support teaching ID in schools after they'd completed a unit on natural versus supernatural explanations—even though this unit never specifically mentioned evolution. Learning about "methodological naturalism" didn't change how much weight students gave to the principle of fairness when they formed their opinions about ID, but it did lead them to give more weight to scientific values.

Looking beyond experiments with college students, the evidence suggests a mixed picture on whether public perceptions of evolution reflect news media habits. When Aaron Veenstra and his colleagues analyzed survey data from 2011, they found no sign that conservative news media (Fox News and conservative talk radio) or liberal news media (MSNBC and progressive talk radio) influenced beliefs about the topic.[67] However, respondents who followed Christian television or radio were particularly likely to disbelieve in evolution.

To take another snapshot of how Americans' news habits line up with their views on evolution, we asked our 2016 survey respondents whether they believed that "humans and other forms of life have evolved over time." The patterns here were less dramatic than the ones for climate change (Figure 5.2). For example, the gap between CNN viewers (92%) and non-viewers (82%) on evolution was ten points, and the one between MSNBC viewers (92%) and non-viewers (83%) was nine points—not trivial, but half the size of the divides on global warming. Likewise, a 9-point gap separated Fox News viewers (76%) and non-viewers (85%)—far short of the 26-point difference on climate change. Furthermore, most of the relationships between media use and perceptions of evolution faded away when we controlled for partisanship, ideology, religiosity, and demographics. One likely explanation for the weaker links here is that the "evolution wars" have received less coverage than global warming. When we tallied cable news stories about evolution, creationism, and ID, we found that coverage peaked in 2005 and declined over the next ten years—the reverse of the trend for climate change.

Even so, two clear patterns emerged from our 2016 survey: watching CNN or MSNBC was tied to believing in evolution, and so was learning from science magazines. In each case, the outlets in question may have been "preaching to the choir" on the topic. Most Americans who read magazines such as *Scientific American* and *National Geographic* or tune in to see Bill Nye appear on CNN or

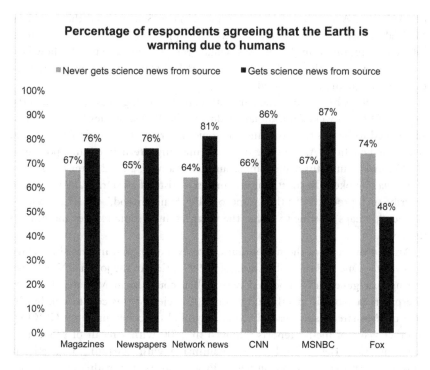

Percentage of respondents agreeing that the Earth is warming due to humans

FIGURE 5.2   Beliefs about evolution, by news media use (Cooperative Congressional Election Survey, 2016)

MSNBC are probably inclined to accept the pro-evolution consensus frames endorsed by scientists. By the same token, many fundamentalist audience members for Christian television and radio presumably belong to the "opposing choir" and are ready to accept the anti-evolution frames promoted by creationists such as Ken Ham.

News about GMOs

The public divide over global warming falls along clear political fault lines: most liberals and Democrats accept the scientific consensus on climate change, whereas most conservatives and Republicans reject it.[68] Similarly, public perceptions of evolution map closely onto a broader split between fundamentalist Christians and non-fundamentalists.[69] By contrast, media framing of GMOs has played out against a murkier social backdrop pitting the promise of increased food production against activists' concerns about health risks, environmental effects, and the practices of agricultural corporations such as Monsanto.[70] Though scientific organizations such as the National Academy of Sciences, the American Association for the Advancement of Science, and the American Medical Association have publicly stated that approved GMO foods are safe to eat, voices ranging from the environmental group Greenpeace to talk show host Dr. Mehmet Oz to blogger Vani "the Food Babe" Hari have argued for restrictions on or labeling of such foods.

News outlets began covering agricultural biotechnology in earnest when the first GMO foods went on the market during the 1990s. Stories from this decade tended to be positive, framing the topic in terms of scientific progress and economic benefits.[71] At the same time, an emerging alternative frame portrayed GMO foods as unnatural and dangerous. For example, a 1992 letter to the *New York Times* invoked the most famous "mad scientist" of all, Dr. Frankenstein, to describe such products. "If they want to sell us Frankenfood," the letter's author argued, "perhaps it's time to gather the villagers, light some torches and head to the castle."[72]

Around the turn of the millennium, two events sparked increased coverage of GMOs.[73] One was the publication of a 1999 article in the journal *Nature* suggesting that genetically modified corn pollen could harm Monarch butterfly caterpillars (a more-in depth study from 2001 rejected this conclusion, but by then the Monarch had already become a symbol for concerns about GMOs).[74] Another wave of media reports followed in 2000 with a recall of products—including Taco Bell taco shells—contaminated with StarLink, a genetically modified corn that the Environmental Protection Agency hadn't approved for human consumption.[75]

Given how these stories highlighted potential risks from agricultural biotechnology, it's not surprising that news coverage from the late 1990s and early 2000s presented a mixed picture of GMOs. In their study of television news, Mary Nucci and Robert Kubey observed a roughly even split between pro- and anti-GMO messages.[76] Likewise, Toby Ten Eyck and Melissa Willement found that newspapers tended to frame debates over agricultural biotechnology in ambivalent ways.[77] Matt Nisbet and Mike Huge, in turn, found that newspaper stories about GMOs presented a range of frames, including technical ones revolving around research and policy as well as more dramatic ones involving strategy, uncertainty, and ethics.[78]

Since the early 2000s, media coverage of GMOs has relied on the same opportunity-based and risk-based frames even as new events have developed.[79] Take the July 13, 2011, ABC News story about GMOs and Monarch butterflies.[80] It begins with a runaway science frame, citing a scientist who calls the use of genetically modified crops and milkweed-killing herbicides such as Roundup a "leading culprit" in declining Monarch butterfly populations. ABC News then balances this frame with progress and economic benefits frames, observing that agricultural biotech has been a "boon to farmers." The story also presents an uncertainty frame by quoting Monsanto (the producer of Roundup) as saying that the evidence about "whether and how agriculture ...affects Monarch population biology" is "disputed" and "still evolving."

In addition to reporting on new studies, media outlets have covered the ongoing debate over laws that would require companies to label GMO foods. One high-profile campaign on the issue revolved around a 2012 California ballot initiative (Proposition 37) to mandate such labels. The initiative's opponents framed GMO foods as safe and economically beneficial while extolling voluntary labeling over mandatory labeling; by contrast, the initiative's supporters framed GMOs in terms of health dangers and the proposition as promoting freedom of information.[81] Though initial public polling found a majority in favor of Proposition 37, California voters ultimately rejected it. Similar proposals failed at the ballot box in Washington state in 2013 and in Colorado and Oregon in 2014.

Two years later, Vermont became the first state to pass a law requiring labels for GMOs in food. Shortly thereafter, however, President Barack Obama signed a law creating a federal GMO labeling system that overturned any state-level requirements. In 2018, the US Department of Agriculture laid out rules—to take effect in 2022—for labeling foods as "bioengineered," rather than "GMO" or "genetically modified." The agricultural industry praised the rules, while critics framed them as an attempt to hide GMOs behind unfamiliar terminology.[82] During all this, news stories about GMOs featured a mixed bag of frames, including criticisms of agricultural companies for putting "the essential ingredients of Agent Orange" in genetically modified corn (a February 12, 2015 episode of MSNBC's *The Ed Show*), criticisms of the restaurant chain Chipotle for touting its GMO-free menu while serving food tainted with E. coli bacteria (a December 8, 2015 episode of *The Five* on Fox News), and celebrity arguments for GMO labeling (an April 23, 2015 interview with Dr. Oz on *CNN Tonight*).

As to whether news coverage shapes public perceptions of GMOs, early studies yielded mixed results. When Susanna Priest analyzed survey data from 2000, she found little evidence of media influence on opinions about agricultural biotechnology.[83] Likewise, Dominique Brossard and Matt Nisbet's look at 2001 survey data yielded no evidence of direct links between respondents' news habits and their views on the issue.[84] On the other hand, John Besley and James Shanahan's analysis of 2003 survey data found that paying attention to television science news was associated with greater support for agricultural biotechnology.[85] More recently, Ivanka Pjesivac and her colleagues found that experimental participants who read an opportunity-framed story came away with more positive attitudes toward GMOs while those who read a risk-framed story shifted toward more negative views.[86]

The results from our 2016 survey suggest that public perceptions of the issue reflect news reading habits more than news viewing habits. Learning about

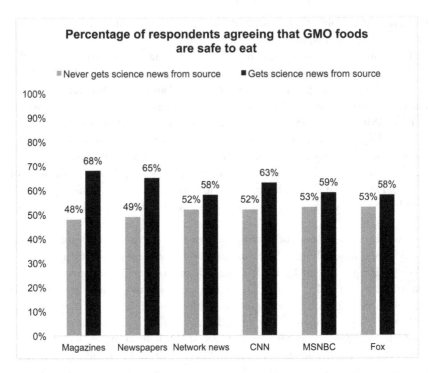

FIGURE 5.3 Beliefs about GMO foods, by news media use (Cooperative Congressional Election Survey, 2016)

science from print sources was linked to believing that it's "generally safe to eat genetically modified foods" (Figure 5.3), with double-digit gaps separating science magazine readers (68%) from non-readers (48%) and newspaper readers (65%) from non-readers (49%). Both patterns remained clear when we controlled for other forms of media use along with demographics, political attitudes, and religiosity. Meanwhile, the differences for network news (58% for viewers versus 52% for non-viewers), CNN (63% versus 52%), MSNBC (59% versus 53%), and Fox News (58% versus 53%) were not only smaller but faded away once we took other factors into account.

The patterns for newspapers and science magazines could reflect framing that emphasizes social progress and economic benefits from GMOs, along with the scientific community's position that genetically modified foods are safe to eat. More broadly, however, public perceptions here mirror the news media's ambivalent framing. Just as our survey found an almost equal split between respondents who saw genetically modified foods as unsafe and those who saw them as safe, more recent surveys have found similar divides.[87] So long as news coverage of the issue highlights both opportunities and risks from GMOs, it may give audience members on both sides of the issue building blocks for reinforcing their existing views.

News about Vaccines

In 1998—the year after the negotiation of the Kyoto climate treaty and the year before the *Nature* article on GMOs and Monarch butterflies—a prestigious medical journal, *The Lancet*, published a study by British doctor Andrew Wakefield that linked the MMR (measles, mumps, and rubella) vaccine to autism. Though Wakefield's study was based on a sample of only 12 self-selected cases and went against the consensus within the scientific community, it played a key role in popularizing the notion that vaccines can cause autism.

To be sure, public concerns about immunizations didn't originate with Wakefield's study. As far back as 1721, Bostonians were arguing among themselves over inoculating citizens against smallpox—so fiercely, in fact, that someone threw a bomb through the window of Cotton Mather, the leading advocate of the immunization campaign. During the 19th century, anti-vaccination leagues formed in various locales around the United States, and in 1982 a television documentary, *DPT: Vaccine Roulette*, fanned concerns about health risks from the vaccine for diphtheria, pertussis, and tetanus.[88]

The publication of Wakefield's study helped launch a new phase of the anti-vaccination movement—as well as a new round of media reports about health concerns involving vaccines.[89] Some of these stories balanced frames emphasizing the benefits of immunization with frames promoted by anti-vaccination activists, including Pandora's box frames about unintended health risks and morality frames about the evils of vaccinations.[90] For example, an October 5, 1999, CNN segment featured parents who argued that immunizations triggered their children's autism. One parent described how his son "got the MMR vaccine, the chicken pox vaccine and oral polio, and within days [he] was a completely different child. I mean, we lost him completely." The segment also featured comments from doctors who rejected Wakefield's conclusions and pointed out flaws in his research, but it ended with a sound bite from a non-medical source, US Representative Dan Burton:

> At the New Jersey conference, Stop Autism Now, there were 1,200 parents there, and they asked the question, "Do you believe the autism of your child was caused by vaccines or related to it?"—750 raised their hands. And across the country, there's been an explosion of autism. We need to find out why.

Thus, CNN gave the last word to someone suggesting a link between vaccines and autism.

Nor were television news programs the only outlets that covered the "autism-vaccine controversy." The number of articles that major newspapers ran about the topic increased from a handful in 1998–2000 to a peak in 2004.[91] A majority of these stories debunked vaccine-autism links, but more than a third of them framed such links as "possible, plausible, or probable."[92]

A few years later, the anti-vaccine movement gained a prominent new spokesperson: actor Jenny McCarthy, who visited *The Oprah Winfrey Show* on September 8, 2007, to frame the MMR vaccine as the cause of her son Evan's autism:

WINFREY: So what do you think triggered the autism? I know you have a theory.

McCARTHY: I do have a theory … Right before his MMR shot, I said to the doctor, I have a very bad feeling about this shot. This is the autism shot, isn't it? And he said, "No, that is ridiculous. It is a mother's desperate attempt to blame something." And he swore at me … And not soon thereafter, I noticed that change in the pictures: Boom! Soul, gone from his eyes.

Winfrey noted that the CDC (Centers for Disease Control and Prevention) rejects any vaccine-autism link but then allowed McCarthy to reply, "My science is named Evan."

In the following years, McCarthy continued to act as a public face for the anti-vaccination—or, as she framed it, "pro-safe vaccine"—movement and as a champion for Wakefield.[93] Given this, ABC's decision in 2013 to hire her as a co-host of the talk show *The View* provoked strong condemnations from doctors and scientists. Yet news coverage of McCarthy's hiring often used a "he said, she said" frame in presenting her claims about vaccines along with the medical community's responses.[94]

As for Wakefield's original study, it was ultimately revealed to be a medical fraud. After its publication, investigators discovered that he had manipulated data, violated ethical standards for research, and failed to disclose financial conflicts of interest.[95] In 2010, *The Lancet* formally retracted his article.[96] That same year, the United Kingdom's General Medical Council revoked his medical credentials for professional misconduct.[97] Still, none of this deterred Wakefield from promoting his views on vaccines. He even directed a 2016 documentary, titled *Vaxxed*, that portrayed the CDC as covering up links between vaccines and autism.

Over the past decade, news organizations have also emphasized controversies surrounding political candidates' statements about vaccines. In September 2011, for example, campaign reporters highlighted Republican presidential candidate Michele Bachmann's suggestion that the human papilloma virus (HPV) vaccine puts "little children's lives at risk" and might cause "mental retardation" (the American Academy of Pediatrics swiftly rebutted her claims).[98] Similarly, Republican presidential candidate Donald Trump stirred media controversy by claiming during a September 17, 2015, CNN debate that vaccines can cause autism. "You take this little beautiful baby," he said, "and you pump — I mean, it looks just like it is meant for a horse, not for a child, and we had so many instances, people that work for me, just the other day, two years old, beautiful child went to have the vaccine and came back and a week later got a tremendous fever, got very, very sick, now is autistic."

To capture the effects of media messages on public perceptions of vaccines, Graham Dixon and Christopher Clarke conducted an experiment in which they randomly assigned participants to read different versions of a news story about the topic.[99] When the story balanced competing claims about the MMR vaccine and autism, readers came away less certain that the vaccine was safe and more likely to think that experts were divided on the subject. These findings suggest that the media frames promoted by anti-vaccination activists can sometimes sway audience members' perceptions.

Looking at the bigger picture, Aaron Veenstra and his colleagues found little evidence that either conservative news media or liberal news media shaped perceptions of whether vaccinations can cause autism.[100] For the most part, the results of our 2016 survey told a similar story. We asked respondents whether they believed that "vaccines for diseases such as measles, mumps, and rubella are safe for healthy children." A large majority agreed, regardless of where they learned, or didn't learn, about science (Figure 5.4). Fully 96% of science magazine and newspaper readers believed the statement, but so did 89% of non-readers. If anything, the differences for television news viewers and non-viewers were even smaller: 95% versus 89% for broadcast news, 95% versus 90% for CNN, 94% versus 91% for MSNBC, and 94% versus 91% for Fox News. However, two of these gaps persisted when we controlled for other factors: science magazine readers were more likely than non-readers to think vaccines are safe, and broadcast network news viewers were marginally more likely than non-viewers to think so.

Overall, a clear majority of the public perceives the benefits from vaccines as outweighing any risks from them. At the same time, news framing that emphasizes the social progress from immunizations and the pro-vaccine consensus among experts may have only a limited impact in changing the minds of anti-vaxxers—many of whom are unlikely to read, let alone trust, sources such as *Scientific American*. Moreover, clashing frames for the issue seem unlikely to disappear anytime soon. For example, consider the March 3, 2019, NBC News story about Ethan Lindenberger, a teenager who got vaccinated against his mother's wishes. It quotes him as saying that claims about

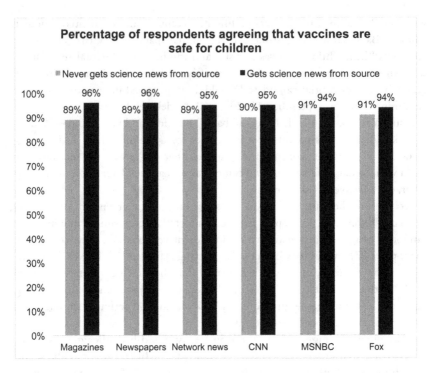

FIGURE 5.4 Beliefs about vaccines, by news media use (Cooperative Congressional Election Survey, 2016)

vaccines causing "autism, brain damage, and other complications [have] been largely debunked by the scientific community." It then quotes his mother as saying that not immunizing her son "was the best way to protect him and keep him safe."[101]

In 2020, the COVID-19 pandemic brought a spike of vaccine-related coverage that featured a familiar set of competing frames along with a sharp political divide. As researchers raced to develop vaccines for the virus, media outlets followed with stories casting them in terms of potential benefits to society (a social progress frame), potential side effects (a Pandora's box frame), questions about efficacy (an uncertainty frame), and the government's role in their development and distribution (a public accountability frame). News outlets also covered—and often attempted to debunk—misinformation about COVID vaccines, including conspiracy theories that they would be used to track and control people.[102]

In an echo of the pattern for climate change, the case of COVID-19 further highlights how science coverage can become entangled in political polarization.

Even in the early stages of the pandemic, Sol Hart and his colleagues found that newspapers and television news stories presented polarized coverage and frequently featured views from politicians along with scientific sources.[103] Given this, it's no surprise that public perceptions of COVID-19 quickly split along party lines. By late 2020, Gallup polling found that 75% of Democrats but only 50% of Republicans were willing to receive a vaccine for the virus.[104]

Other Issues to Frame, Other Platforms for Framing

News framing matters for how audience members see science, even though its effects aren't the same across every issue or every outlet. For some issues—such as climate change and COVID-19 vaccines—cable news networks can play a key role in reinforcing polarized views among audience members. For other issues—such as evolution and stem cell research—the effects of news framing may be more modest and concentrated among specific audiences, such as the ones for science media or religious media.

News framing is also a double-edged sword: media messages can lead audience members toward accepting *or* rejecting the scientific consensus on issues. In many cases, frames in news coverage can promote greater agreement with scientists. Furthermore, messages that teach scientific reasoning may help bridge gaps between audience members and scientists. At the same time, framing can also serve as a tool for challenging what the scientific community says. The dismissive frames often featured on Fox News can undermine belief in climate change, just as the fairness frames promoted by organizations such as the Discovery Institute can promote support for teaching ID in schools. Similarly, risk frames can fuel negative views toward GMOs, and "balanced" framing can undermine perceptions of vaccine safety.

Looking forward, it's a safe bet that the news media will use the frames we've discussed to cover other emerging scientific issues. Take CRISPR (clustered regularly interspaced short palindromic repeats), a recently developed method that allows for relatively cheap and easy editing of genes. Coverage of this technology went from only a few stories in 2015 to hundreds of stories by the end of 2018, driven by events such as a high-stakes patent battle, the release of a movie (*Rampage*) featuring Dwayne "the Rock" Johnson fighting CRISPR-modified creatures, and Chinese scientist He Jiankui's announcement that he'd used CRISPR to edit the genes of two newborn twin girls. When D. J. McCauley analyzed these stories, she found that they used almost every frame we've discussed, including progress frames emphasizing medical and agricultural applications, ethics frames revolving around the pitfalls of editing human DNA, runaway science frames dramatizing the dangers of "rogue science," and strategy frames highlighting maneuvers in patent battles.[105] Just as news framing of stem cell research influenced public perceptions a decade and a half ago, these frames for CRISPR may do the same—particularly if news outlets continue to devote increasing attention to the topic.

In weighing the effects of such framing, however, it's important to keep in mind not only the media practitioners who present frames but also their intended audiences. Members of the public aren't empty vessels waiting to be filled with frames; instead, they're active, and often critical, participants in the framing process. Some audience members may accept a given news frame for a science-related issue, particularly when that frame resonates with their own beliefs as well as broader cultural narratives. Other members of the public may reject the same frame or even reinterpret it to bolster their preexisting views. Still other people will never receive the frame in the first place because they don't follow the sources that present it, either due to selective exposure based on their prior beliefs or lack of interest in traditional science news.

Which brings up another key point: many people get science news from sources besides magazines, newspapers, and traditional television news. For example, jokes such as the ones Stephen Colbert made about global warming on *The Colbert Report* could shape what audience members think about science-related issues. As it happens, he's not the only late-night television comedian to talk about the topics we've looked at in this chapter—and more besides.

Notes

1 Pew Research Center, "Public and scientists' views on science and society," Pew Research Center, Jan. 29, 2015, www.pewinternet.org/2015/01/29/public-and-scientists-views-on-science-and-society.

2 Goffman, Erving, *Frame analysis: An essay on the organization of experience*, Harvard University Press, 1974.

3 Tversky, Amos, and Daniel Kahneman, "The framing of decisions and the psychology of choice," *Science* 211, no. 4481 (1981): 453–458.

4 Gamson, William A., and Andre Modigliani, "Media discourse and public opinion on nuclear power: A constructionist approach," *American Journal of Sociology* 95, no. 1 (1989): 1–37.

5 Entman, Robert M., "Framing: Toward clarification of a fractured paradigm," *Journal of Communication* 43, no. 4 (1993): 51–58.

6 Iyengar, Shanto, *Is anyone responsible? How television frames political issues*, University of Chicago Press, 1994.

7 Hart, Philip Solomon, "One or many? The influence of episodic and thematic climate change frames on policy preferences and individual behavior change," *Science Communication* 33, no. 1 (2011): 28–51.

8 Muter, Bret A., Meredith L. Gore, Katie S. Gledhill, Christopher Lamont, and Charlie Huveneers, "Australian and US news media portrayal of sharks and their conservation," *Conservation Biology* 27, no. 1 (2013): 187–196.

9 Nisbet, Matthew C., "Framing science: A new paradigm in public engagement," in *Understanding science: New agendas in science communication*, eds. LeeAnn Kahlor and Patricia A. Stout, Routledge, 2009, 40–67.

10 Nisbet, Matthew C., Dominique Brossard, and Adrianne Kroepsch, "Framing science: The stem cell controversy in an age of press/politics," *Harvard International Journal of Press/Politics* 8, no. 2 (2003): 36–70.

11 Colarossi, Jessica, "Comparing how media around the world frames coronavirus news," *The Brink: Pioneering Research from Boston University*, June 25, 2020, www.bu.edu/articles/2020/comparing-how-media-around-the-world-frames-coronavirus-news.

12 Nelkin, Dorothy, *Selling science: How the press covers science and technology*, rev. edition, W. H. Freeman, 1995.

13 Entman, Robert M., *Democracy without citizens: Media and the decay of American politics*, Oxford University Press, 1990.

14 Dearing, James W., "Newspaper coverage of maverick science: Creating controversy through balancing," *Public Understanding of Science* 4, no. 4 (1995): 341–361.

15 Brewer, Paul R., David Wise, and Barbara L. Ley, "Chemical controversy: Canadian and US news coverage of the scientific debate about bisphenol A," *Environmental Communication* 8, no. 1 (2014): 21–38.

16 Mooney, Chris, "Blinded by science: How 'balanced' coverage lets the scientific fringe hijack reality," *Columbia Journalism Review* 43, no. 4 (2004): 26–36.

17 Overby, Dennis, "Gauging a collider's odds of creating a black hole," *New York Times*, Apr. 15, 2008, www.nytimes.com/2008/04/15/science/15risk.html; Overby, Dennis, "The collider, the particle and a theory about fate," *New York Times*, Oct. 13, 2009, www.nytimes.com/2009/10/13/science/space/13lhc.html.

18 Obozintsev, Lucy, "From Skynet to Siri: An exploration of the nature and effects of media coverage of artificial intelligence," MA thesis, University of Delaware, 2018.

19 Fitzgerald, Scott T., and Beth A. Rubin, "Risk society, media, and power: The case of nanotechnology," *Sociological Spectrum* 30, no. 4 (2010): 367–402.

20 Antilla, Liisa, "Climate of scepticism: US newspaper coverage of the science of climate change," *Global Environmental Change* 15, no. 4 (2005): 338–352; Dunlap, Riley E., Aaron M. McCright, and Jerrod H. Yarosh, "The political divide on climate change: Partisan polarization widens in the US," *Environment: Science and Policy for Sustainable Development* 58, no. 5 (2016): 4–23.

21 Mooney, Chris, and Matthew C. Nisbet, "Undoing Darwin," *Columbia Journalism Review* 44, no. 3 (2005): 30–39.

22 Crawley, Catherine E., "Localized debates of agricultural biotechnology in community newspapers: A quantitative content analysis of media frames and sources," *Science Communication* 28, no. 3 (2007): 314–346; Nisbet, Matthew C., and Mike Huge, "Where do science debates come from? Understanding attention cycles and framing," in *The media, the public and agricultural biotechnology*, eds. Dominque Brossard, Thomas C. Nesbitt, and James Shanahan, CABI, 2002, 193–230.

23 Nyhan, Brendan, "Why 'he said, she said' is dangerous," *Columbia Journalism Review*, Jul. 16, 2013, https://archives.cjr.org/united_states_project/media_errs_giving_balanced_coverage_to_jenny_mccarthys_discredited_views.php.

24 Nisbet, Matthew C., and Chris Mooney, "Framing science," *Science* 316, no. 5821 (2009): 56.

25 Kavanagh, Etta (ed.), "The risks and advantages of framing science," *Science* 317, no. 5842 (2009): 1168–1170.

26 Hart, "One or many?"

27 Nisbet, Matthew C., and Robert K. Goidel, "Understanding citizen perceptions of science controversy: Bridging the ethnographic—survey research divide," *Public Understanding of science* 16, no. 4 (2007): 421–440.

28 Scheufele, Dietram A., and Bruce V. Lewenstein, "The public and nanotechnology: How citizens make sense of emerging technologies," *Journal of Nanoparticle Research* 7, no. 6 (2005): 659–667.

29 Gamson, William A., *Talking politics*, Cambridge University Press, 1992.

30 Walsh, Katherine Cramer, *Talking about politics: Informal groups and social identity in American life*, University of Chicago Press, 2004.

31 Cobb, Michael D., "Framing effects on public opinion about nanotechnology," *Science Communication* 27, no. 2 (2005): 221–239; Druckman, James N., "On the limits of framing effects: Who can frame?" *Journal of Politics* 63, no. 4 (2001): 1041–1066.

32 Chong, Dennis, and James N. Druckman, "Framing public opinion in competitive democracies," *American Political Science Review* (2007): 637–655; Nisbet, Erik C., P. Sol Hart, Teresa Myers, and Morgan Ellithorpe, "Attitude change in competitive framing environments? Open-/closed-mindedness, framing effects, and climate change," *Journal of Communication* 63, no. 4 (2013): 766–785.

33 Kahan, Dan M., "Evidence-based science filmmaking initiative: Study no. 1," Jan. 11, 2016, https://ssrn.com/abstract=2713563.

34 Stroud, Natalie Jomini, "Polarization and partisan selective exposure," *Journal of communication* 60, no. 3 (2010): 556–576.

35 Bolsen, Toby, and Matthew A. Shapiro, "The US news media, polarization on climate change, and pathways to effective communication," *Environmental Communication* 12, no. 2 (2018): 149–163; McCright, Aaron M., and Riley E. Dunlap, "Defeating Kyoto: The conservative movement's impact on US climate change policy," *Social Problems* 50, no. 3 (2003): 348–373.

36 Trumbo, Craig, "Constructing climate change: Claims and frames in US news coverage of an environmental issue," *Public understanding of science* 5, no. 3 (1996): 269–284.

37 Boykoff, Maxwell, T., Meaghan Daly, Lucy McAllister, Marisa McNatt, Ami Nacu-Schmidt, David Oonk, and Olivia Pearman, "United States coverage of climate change or global warming, 2000–2018," Center for Science and Technology Policy Research, Cooperative Institute for Research in Environmental Sciences, University of Colorado, Dec. 31, 2018.

38 Bolsen and Shapiro, "The US news media"; Hart, P. Sol, and Lauren Feldman, "Threat without efficacy? Climate change on US network news," *Science Communication* 36, no. 3 (2014): 325–351; Feldman, Lauren, P. Sol Hart, and Tijana Milosevic, "Polarizing news? Representations of threat and efficacy in leading US newspapers' coverage of climate change," *Public Understanding of Science* 26, no. 4 (2017): 481–497; Nisbet, Matthew C., "Communicating climate change: Why frames matter for public engagement," *Environment: Science and policy for sustainable development* 51, no. 2 (2009): 12–23.

39 Cook, John, Dana Nuccitelli, Sarah A. Green, Mark Richardson, Baerbel Winkler, Rob Painting, Robert Way, Peter Jacobs, and Andrew Skuce, "Quantifying the consensus on anthropogenic global warming in the scientific literature," *Environmental Research Letters* 8, no. 2 (2013): 024024; Oreskes, Naomi, "The scientific consensus on climate change," *Science* 306, no. 5702 (2004): 1686–1686.

40 Boykoff, Maxwell T, "Lost in translation? United States television news coverage of anthropogenic climate change, 1995–2004," *Climatic Change* 86, no. 1–2 (2008): 1–11; Boykoff, Maxwell T., and Jules M. Boykoff, "Balance as bias: Global warming and the US prestige press," *Global environmental change* 14, no. 2 (2004): 125–136.

41 Boykoff and Boykoff, "Balance as bias," 126.

42 Boykoff, Maxwell T, "Flogging a dead norm? Newspaper coverage of anthropogenic climate change in the United States and United Kingdom from 2003 to 2006," *Area* 39, no. 4 (2007): 470–481.

43 Nisbet, Matthew C., *Climate shift: Clear vision for the next decade of public debate,* American University School of Communication, 2011.

44 Feldman, Lauren, Edward W. Maibach, Connie Roser-Renouf, and Anthony Leiserowitz, "Climate on cable: The nature and impact of global warming coverage on Fox News, CNN, and MSNBC," *International Journal of Press/Politics* 17, no. 1 (2012): 3–31.

45   Ahern, Lee, and Melanie Formentin, "More is less: Global warming news values on Fox compared to other US broadcast news outlets," *Electronic News* 10, no. 1 (2016): 45–65.

46   Myers, Teresa A., Matthew C. Nisbet, Edward W. Maibach, and Anthony A. Leiserowitz, "A public health frame arouses hopeful emotions about climate change," *Climatic Change* 113, no. 3–4 (2012): 1105–1112.

47   Feinberg, Matthew, and Robb Willer, "The moral roots of environmental attitudes," *Psychological Science* 24, no. 1 (2013): 56–62.

48   Corbett, Julia B., and Jessica L. Durfee, "Testing public (un) certainty of science: Media representations of global warming," *Science Communication* 26, no. 2 (2004): 129–151.

49   Hart, P. Sol, and Erik C. Nisbet, "Boomerang effects in science communication: How motivated reasoning and identity cues amplify opinion polarization about climate mitigation policies," *Communication Research* 39, no. 6 (2012): 701–723.

50   Hart, P. Sol, Erik C. Nisbet, and Teresa A. Myers, "Public attention to science and political news and support for climate change mitigation," *Nature Climate Change* 5, no. 6 (2015): 541–545.

51   Feldman et al., "Climate on cable."

52   Feldman, Lauren, Teresa A. Myers, Jay D. Hmielowski, and Anthony Leiserowitz, "The mutual reinforcement of media selectivity and effects: Testing the reinforcing spirals framework in the context of global warming," *Journal of Communication* 64, no. 4 (2014): 590–611.

53   McCarthy, Joe, "A major national climate report came out. Then the deniers got on TV," *The Weather Channel*, Nov. 29, 2018, https://weather.com/science/environment/news/2018-11-29-climate-deniers-found-themselves-on-tv.

54   *State of the Union*, CNN, Nov. 25, 2018.

55   Pew Research Center, "Public and scientists' views"; Plutzer, Eric, and Michael Berkman, "Trends: Evolution, creationism, and the teaching of human origins in schools," *Public Opinion Quarterly* 72, no. 3 (2008): 540–553.

56   Brod, Donald F., "The Scopes trial: A look at press coverage after forty years," *Journalism Quarterly* 42, no. 2 (1965): 219–226.

57   Taylor, Charles Alan, and Celeste Michelle Condit, "Objectivity and elites: A creation science trial," *Critical Studies in Media Communication* 5, no. 4 (1988): 293–312.

58   Slevin, Peter, "Teachers, scientists vow to fight challenge to evolution," *Washington Post*, May 5, 2005, www.washingtonpost.com/archive/politics/2005/05/05/teachers-scientists-vow-to-fight-challenge-to-evolution/bb1d77ea-8fb9-4d1d-8e86-0fa01c73bb70/.

59   Wallis, Claudia, "The evolution wars," *TIME*, Aug. 7, 2005; Forrest, Barbara, and Paul R. Gross, *Creationism's Trojan horse: The wedge of intelligent design*, Oxford University Press, 2007.

60   Mooney and Nisbet, "Undoing Darwin."

61   Rosenhouse, Jason, and Glenn Branch, "Media coverage of 'intelligent design,'" *BioScience* 56, no. 3 (2006): 247–252.

62   Mooney and Nisbet, "Undoing Darwin."

63   Grimm, Joshua, "'Teach the controversy: The relationship between sources and frames in reporting the intelligent design debate," *Science Communication* 31, no. 2 (2009): 167–186.

64   Matzke, Nicholas J., "The evolution of antievolution policies after Kitzmiller versus Dover," *Science* 351, no. 6268 (2016): 28–30.

65   Nye, Bill, "Bill Nye's take on the Nye-Ham debate," *Skeptical Inquirer*, May/June 2014, www.csicop.org/si/show/bill_nyes_take_on_the_nye-ham_debate.

66   Nelson, Thomas E., Dana E. Wittmer, and Dustin Carnahan, "Should science class be fair? Frames and values in the evolution debate," *Political Communication* 32, no. 4 (2015): 625–647.

67   Veenstra, Aaron S., Mohammad Delwar Hossain, and Benjamin A. Lyons, "Partisan media and discussion as enhancers of the belief gap," *Mass Communication and Society* 17, no. 6 (2014): 874–897.

68   Funk, Cary, Brian Kennedy, Meg Hefferon and Mark Strauss, "Majorities see government efforts to protect the environment as insufficient," Pew Research Center, May 14, 2018, www.pewresearch.org/science/2018/05/14/majorities-see-government-efforts-to-protect-the-environment-as-insufficient/.

69   Funk, Cary, "How highly religious Americans view evolution depends on how they're asked about it," Pew Research Center, Feb. 6, 2019, www.pewresearch.org/fact-tank/2019/02/06/how-highly-religious-americans-view-evolution-depends-on-how-theyre-asked-about-it/.

70   Funk, Cary and Brian Kennedy, "The new food fights: U.S. public divides over food science," Pew Research Center, Dec. 1, 2016, www.pewresearch.org/science/2016/12/01/public-opinion-about-genetically-modified-foods-and-trust-in-scientists-connected-with-these-foods/.

71   Nisbet, Matthew C., and Bruce V. Lewenstein, "Biotechnology and the American media: The policy process and the elite press, 1970 to 1999," *Science Communication* 23, no. 4 (2002): 359–391.

72   Lewis, Paul, "Mutant foods create risks we can't yet guess; since Mary Shelley," *New York Times*, June 16, 1992, www.nytimes.com/1992/06/16/opinion/l-mutant-foods-create-risks-we-can-t-yet-guess-since-mary-shelley-332792.html.

73   Marks, Leonie A., Nicholas G. Kalaitzandonakes, Kevin Allison, and Ludmila Zakharova, "Media coverage of agrobiotechnology: Did the butterfly have an effect?" *Journal of Agribusiness* 21, no. 345–2016–15206 (2003): 1–20; Nucci, Mary L., and Robert Kubey, "'We begin tonight with fruits and vegetables': Genetically modified food on the evening news 1980–2003," *Science Communication* 29, no. 2 (2007): 147–176.

74   Losey, John E., Linda S. Rayor, and Maureen E. Carter, "Transgenic pollen harms monarch larvae," *Nature* 399, no. 6733 (1999): 214–214; Sears, Mark K., Richard L. Hellmich, Diane E. Stanley-Horn, Karen S. Oberhauser, John M. Pleasants, Heather R. Mattila, Blair D. Siegfried, and Galen P. Dively, "Impact of Bt corn pollen on monarch butterfly populations: A risk assessment," *Proceedings of the National Academy of Sciences* 98, no. 21 (2001): 11937–11942.

75   Nisbet and Huge, "Where do science debates come from?"

76   Nucci and Kubey, "We begin tonight."

77   Eyck, Toby A. Ten, and Melissa Williment, "The national media and things genetic: Coverage in the *New York Times* (1971–2001) and the *Washington Post* (1977–2001)," *Science Communication* 25, no. 2 (2003): 129–152.

78   Nisbet and Huge, "Where do science debates come from?"

79   Pjesivac, Ivanka, Marlit A. Hayslett, and Matthew T. Binford, "To eat or not to eat: Framing of GMOs in American media and its effects on attitudes and behaviors," *Science Communication* 42, no. 6 (2020): 747–775.

80   Potter, Ned, "Are monarch butterflies threatened by genetically modified crops?" ABC News, July 13, 2011, https://abcnews.go.com/Technology/monarch-butterflies-genetically-modified-gm-crops/story?id=14057436.

81   Zilberman, David, Scott Kaplan, Eunice Kim, and Gina Waterfield, "Lessons from the California GM labelling proposition on the state of crop biotechnology," in *Handbook on Agriculture, Biotechnology and Development*, eds. Stuart J. Smyth, Peter W.B. Phillips, and David Castle, Edward Elgar Publishing, 2014, 538–549; Krause, Amber, Courtney Meyers, Erica Irlbeck, and Todd Chambers, "What side are you on? An examination of the persuasive message factors in proposition 37 videos on YouTube," *Journal of Applied Communications* 100, no. 3 (2016): 68–83.

82   Prentice, Chris, and Jonathan Otis, "USDA outlines first-ever rule for GMO labeling, sees implementation in 2020," Reuters, Dec. 12, 2018,

www.reuters.com/article/us-usa-gmo-labeling/usda-outlines-first-ever-rule-for-gmo-labeling-sees-implementation-in-2020-idUSKCN1OJ2TF; Kennedy, Merrit, "USDA unveils prototypes for GMO food labels, and they're ... confusing," NPR, May 19, 2018, www.npr.org/sections/thesalt/2018/05/19/612063389/usda-unveils-prototypes-for-gmo-food-labels-and-theyre-confusing.

83 Priest, Susanna Hornig, "Misplaced faith: Communication variables as predictors of encouragement for biotechnology development," *Science Communication* 23, no. 2 (2001): 97–110.

84 Brossard, Dominique, and Matthew C. Nisbet, "Deference to scientific authority among a low information public: Understanding US opinion on agricultural biotechnology," *International Journal of Public Opinion Research* 19, no. 1 (2007): 24–52.

85 Besley, John C., and James Shanahan, "Media attention and exposure in relation to support for agricultural biotechnology," *Science Communication* 26, no. 4 (2005): 347–367.

86 Pjesivac et al., "To eat or not to eat."

87 Kennedy, Brian, Meg Hefferon, and Cary Funk, "Americans are narrowly divided over health effects of genetically modified foods," Pew Research Center, Nov. 19, 2018, www.pewresearch.org/fact-tank/2018/11/19/americans-are-narrowly-divided-over-health-effects-of-genetically-modified-foods/.

88 Novak, Sara, "The long history of America's anti-vaccination movement," *Discover*, Nov. 26, 2018, http://discovermagazine.com/2018/dec/fostering-fear.

89 Clarke, Christopher E., "A question of balance: The autism-vaccine controversy in the British and American elite press.," *Science Communication* 30, no. 1 (2008): 77–107.

90 Kata, Anna, "A postmodern Pandora's box: Anti-vaccination misinformation on the Internet," *Vaccine* 28, no. 7 (2010): 1709–1716.

91 Clarke, "A question of balance."

92 Clarke, "A question of balance," 90.

93 "Jenny McCarthy: ''We're not an anti-vaccine movement ... we're pro-safe vaccine,''" *Frontline*, Mar. 23, 2015, www.pbs.org/wgbh/frontline/article/jenny-mccarthy-were-not-an-anti-vaccine-movement-were-pro-safe-vaccine/.

94 Nyhan, "Why 'he said, she said' is dangerous."

95 Belluz, Julia, "Research fraud catalyzed the anti-vaccination movement. Let's not repeat history," *Vox*, Feb. 27, 2018, www.vox.com/2018/2/27/17057990/andrew-wakefield-vaccines-autism-study.

96 Harris, Gardiner, "Journal retracts 1998 paper linking autism to vaccines," *New York Times*, Feb. 3, 2010, www.nytimes.com/2010/02/03/health/research/03lancet.html.

97 Triggle, Nick, "MMR doctor struck from register," BBC News, May 24, 2010, http://news.bbc.co.uk/2/hi/health/8695267.stm.

98 Holan, Angie Drobnic and Louis Jacobson, "Michele Bachmann says HPV vaccine can cause mental retardation," PolitiFact, Sept. 16, 2011, www.politifact.com/truth-o-meter/statements/2011/sep/16/michele-bachmann/bachmann-hpv-vaccine-cause-mental-retardation/.

99 Dixon, Graham N., and Christopher E. Clarke, "Heightening uncertainty around certain science: Media coverage, false balance, and the autism-vaccine controversy," *Science Communication* 35, no. 3 (2013): 358–382; Clarke, Christopher E., Brooke Weberling McKeever, Avery Holton, and Graham N. Dixon, "The influence of weight-of-evidence messages on (vaccine) attitudes: A sequential mediation model," *Journal of Health Communication* 20, no. 11 (2015): 1302–1309.

100 Veenstra et al., "Partisan media."

101 Rosenblatt, Kalhan, "Teen who got all his shots despite anti-vaccine mother to testify before Congress," NBC News, Mar. 3, 2019, www.nbcnews.com/health/health-news/teen-who-got-all-his-shots-despite-anti-vaccine-mother-n978706.

102 Bond, Shannon, "'The perfect storm': How vaccine misinformation spread to the mainstream," NPR, Dec. 10, 2020, www.npr.org/2020/12/10/944408988/the-perfect-storm-how-coronavirus-spread-vaccine-misinformation-to-the-mainstrea.

103 Hart, P. Sol, Sedona Chinn, and Stuart Soroka, "Politicization and polarization in COVID-19 news coverage," *Science Communication* 42, no. 5 (2020): 679–697.

104 Brenan, Megan, "Willingness to get COVID-19 vaccine ticks up to 63% in U.S.," Gallup, Dec. 8, 2020, https://news.gallup.com/poll/327425/willingness-covid-vac-cine-ticks.aspx.

105 McCauley, Darryn Jayne, "Re-writing the genetic code: An exploration of framing, sources, and hype in media coverage of CRISPR," MA thesis, University of Delaware, 2019.

6

LATE-NIGHT SCIENCE

Aasif Mandvi: Science claims it's working to cure disease, save the planet, and solve our greatest mysteries, but what's it really up to? From global warming …

Herman Cain: I don't believe global warming is real.

Mandvi: … to evolution …

Rick Santorum: Absolutely not, I don't believe in that.

Mandvi: … to the HPV vaccine …

Michelle Bachman: Her daughter suffered mental retardation as a result of that vaccine.

Mandvi: … it seems science is up to something.

Rick Perry: There are a substantial number of scientists who have manipulated data so that they will have dollars rolling into their projects.

Mandvi: Could these Republican candidates be right?

—*The Daily Show with Jon Stewart* (October 26, 2011)

John Oliver: If we start thinking that science is à la carte and if you don't like one study, don't worry, another will be along soon, that is what leads people to think that manmade climate change isn't real or that vaccines cause autism, both of which the scientific consensus is pretty clear on.

—*Last Week Tonight with John Oliver* (May 8, 2016)

Science may be a serious endeavor, but it's also funny—so much so that late-night comedy shows have long mined it for laughs. Back in the 1980s, *Late Show* host Johnny Carson imitated astronomer Carl Sagan's fondness for the

DOI: 10.4324/9781003190721-6

word "billions."[1] Similarly, current *Late Show* host Stephen Colbert has joked about scientists studying the plausibility of Spider-Man's wall-climbing abilities (January 22, 2016), the sounds made by "ice ghosts" in Antarctica (October 20, 2018), and the potential for self-driving cars to be used for sex (November 14, 2018).

In some cases, late-night comedians invoke popular stereotypes of scientists as oddball eggheads researching esoteric subjects. For example, an August 5, 2009 segment on the satirical cable television news program *The Daily Show* lampoons two rival primatologists studying how orangutans and chimpanzees are related to humans. During an interview, "correspondent" John Oliver pretends to find the scientists' research mind-numbing:

PROFESSOR JEFFREY SCHWARTZ: Now, what's interesting is if you look at these fossils, human fossils, they have the same orang features of the face.
OLIVER: Does "interesting" mean something different in the scientific field than it does in other life?

Oliver also takes the opportunity to poke fun at the obscurity of scientific publishing:

PROFESSOR TODD DISOTELL: I think the arguments are very easy to counter, and it's going to let me write a counter-paper.
OLIVER: What will he do then—write a counter-paper to your counter-paper?
DISOTELL: Yeah.
OLIVER: Then you'll publish a counter to that, he'll write a counter paper saying that he's right and you're wrong, and no one will read any of them.
DISOTELL: Probably true, unfortunately.

The comedian even jokes about Disotell's mohawk haircut, asking him, "Did chimps, when they lost all their hair … retain a single stripe of hair across the top of their head?"

Yet late-night hosts also use humor to affirm what scientists say and rebut anti-science voices. Jon Stewart was a trendsetter in doing so: during his tenure as host of *The Daily Show* from 1999 to 2015, he and his team often addressed science-related issues. Consider the October 26, 2011 segment titled "Science: What's It Up To?" in which Aasif Mandvi investigates—and mocks—Republican politicians' statements about climate change, evolution, and vaccines. After interviewing political strategist Noelle Nikpour, who claims that scientists are "scamming the American people," the comedian/correspondent visits Columbia biology professor (and Nobel Prize winner) Martin Chalfie's "luxurious palace of science" (actually an ordinary-looking laboratory), calls him a "notorious swindler," and accuses him of running "the oldest grift in the book—the old nematode switcheroo." The segment ends with Mandvi's trip to a science fair, where he warns the young participants against getting "hooked on that grant money."

Stewart's successor, Trevor Noah, has continued this tradition of satirical science commentary, as have several other *Daily Show* alumni-turned-late-night hosts—including Stephen Colbert (of *The Colbert Report* and then *The Late Show*), Larry Wilmore (*The Nightly Show*), John Oliver (*Last Week Tonight*), and Samantha Bee (*Full Frontal*). Likewise, Jimmy Kimmel (*Jimmy Kimmel Live!*), Jimmy Fallon (*The Tonight Show*), and Seth Meyers (*Late Night*) have all addressed science-related issues on their programs.

Late-night television also has a history of providing scientists with platforms for speaking directly to the public. In addition to mimicking Carl Sagan, Johnny Carson regularly invited him to appear as a guest on *The Late Show* during the 1970s and 1980s, giving the astronomer a chance to promote space exploration to audience members.[2] Since then, other hosts have emulated Carson by inviting a stream of scientists for interviews. As a result, late-night viewers have had opportunities to hear guests such as Jane Goodall, Stephen Hawking, Michio Kaku, and Neil deGrasse Tyson discuss their own research along with broader issues involving science and society.

All this matters because late-night television shows draw sizable audiences. In May 2019, for example, *The Late Show*, *The Tonight Show*, and *Jimmy Kimmel Live!* each drew more than two million viewers a night.[3] Meanwhile, segments from *The Daily Show*, *Last Week Tonight*, and *Full Frontal* frequently receive millions, or even tens of millions, of views on video-sharing websites such as YouTube. With their broad reaches, these programs can help promote public engagement with science and influence public perceptions on questions such as whether human activity is causing climate change, whether vaccines are safe for children, and whether genetically modified foods are safe to eat. This is especially true for viewers who don't intentionally seek out news about science.

Satire as a Gateway to Science

As we've seen, traditional media outlets can, and sometimes do, shape perceptions of climate change, evolution, vaccines, and GMOs, among other issues. Yet a "missing audience" for science information pays little, if any, attention to sources such as newspapers, broadcast evening news shows, or cable television news channels—let alone to media outlets that focus on science, such as *NOVA* or *National Geographic*.[4] Furthermore, even news junkies may encounter relatively few stories about science, given the dwindling resources devoted to science journalism over the past few decades.[5] One study by the Project for Excellence in Journalism found that mainstream news sources devoted less than

2% of their coverage to science and technology in 2008, a figure that probably hasn't improved since then.[6]

Compared to traditional news, late-night television sometimes looks like a font of science coverage. Indeed, the same Project for Excellence in Journalism study found that *The Daily Show* devoted twice as much of its 2008 coverage to science and technology stories as did more "serious" outlets. For example, the program's June 18, 2008 episode discusses a space probe's discovery of "mysterious white stuff" on Mars: "We cannot confirm life on Mars," announces Stewart, "but it seems we may be able to confirm *night*life on Mars." Some of the guest interviews from that year also addressed science—including the May 27, 2008 one with physicist Brian Greene, who frames the topic in ways that encourage broader public engagement with it. "The goal is to shift the public's perception of science," he explains. "Many people are intimidated by science … Many people just think about science as what's in the textbook, but science is the active exploration. Science is the greatest of adventure stories."

If anything, *The Daily Show*'s spinoff, *The Colbert Report*, did even more to promote science during its 11-year run. The show's host, a self-proclaimed fan of science, not only discussed the topic in numerous segments but also used his late-night platform to advocate for the space program.[7] In the April 8, 2010 episode of the program, he goes so far as to drop his usual satirical persona— that of a conservative talk host shows in the mold of Bill O'Reilly—and make a direct plea to save NASA's astronaut program.[8] On a more humorous note, he also campaigned for NASA to name a room on its space station after him (the space agency decided not to, but it did name a zero-gravity treadmill after him: the Combined Operational Load-Bearing External Resistance Treadmill, or COLBERT).

This sort of coverage on late-night television can help foster public interest in science.[9] As Lauren Feldman and her colleagues point out, entertainment-oriented "soft news" outlets, including late-night comedy shows, often serve as gateways to engagement with public affairs. In effect, these programs "piggyback" messages about substantive topics on top of the lighthearted content that draws viewers.[10] Looking at survey data from 2008, Feldman and her team found that watching *The Daily Show* and *The Colbert Report* went hand in hand with paying more attention to science and technology. Furthermore, the link between late-night viewing and engagement with science was the strongest among the least educated respondents. In effect, watching Jon Stewart and Stephen Colbert helped bridge a science "engagement gap" between more informed and less informed Americans.

Since his move from *The Colbert Report* to *The Late Show*, Colbert has continued to embrace the role of cheerleader for science, particularly space exploration. For example, a February 24, 2017 segment of his program highlights his visit to NASA headquarters to take part in astronaut training, and a September 21,

2018 segment shows him and Neil deGrasse Tyson test-driving a Mars Rover in Manhattan (at the same time, Colbert has been willing to satirize NASA's more questionable actions—for example, its proposal to raise money through "ads in space" and its cancelation of an all-woman spacewalk due to "the lack of a space-suit in the right size").[11] Nor is Colbert alone in his efforts to promote science: Jimmy Fallon's "Fallonventions" segments on *The Tonight Show* feature children who have built scientific inventions, and Jimmy Kimmel regularly invites "Science Bob" Pflugfelder to perform wacky demonstrations with flying ping-pong balls and exploding pumpkins. By wrapping their messages about science in humor, these hosts can reach audience members who might never tune into *Cosmos* or read *Scientific American*.

Climate as Comedy

Beyond sparking engagement with science, late-night comedy shows may also shape viewers' perceptions of science-related issues. Researchers have found that these programs influence public opinion on subjects ranging from presidential candidates to complicated policy issues such as net neutrality—so why not vaccines or genetically modified foods, as well?[12]

Take the case of climate change, a perennial topic on late-night television for two decades and running. When Lauren Feldman analyzed *Daily Show* coverage from 1999 to 2012 and *Colbert Report* coverage from 2005 to 2012, she found that the programs ran a total of 183 segments dealing with global warming.[13] In terms of *how* they covered the issue, both shows consistently endorsed the scientific consensus that climate change is happening. For example, Stewart's October 4, 2005 monologue invokes the "near-universal consensus" that Arctic ice melting is "due in part to global warming." Similarly, a *Daily Show* segment from February 5, 2007 includes a dubbed "warning" from groundhog Punxsutawney Phil—"There will be cataclysmic climate change of anthropogenic origin. The ice caps will melt at an ever-accelerating rate … I speak the truth!"—along with Stewart's description of the conclusions from the Intergovernmental Panel on Climate Change (IPCC): "To sum up, the report states that global warming is happening and that it's all but certainly a man-made phenomenon."

Furthermore, the two hosts often mocked climate change skeptics. Stewart tended to use sarcasm in doing so, comparing them to the one dentist in five who "recommends sugared gum" in sugar-free gum commercials and ridiculing their attacks on "fat-cat scientists with their easy double-blind study money" (December 14, 2009). Meanwhile, Colbert used his ironic persona to satirize climate change skeptics. In one monologue about the IPCC report, he says, "Sure, this new report sounds horrible, but as long as we can find one scientist who says it's not happening, then it's not happening … because even if 99.9% of experts believe global warming is happening, as long as there's any disagreement, well, all the science isn't in" (February 5, 2007). In another episode, he proclaims, "Nation, no one will ever convince me that global warming is real—just like Craisins. Throw all the science you want at me, Ocean Spray, but a cranberry that's also a raisin? That's the devil's snack" (April 26, 2011).

During the time span Feldman studied, *The Daily Show* and *The Colbert Report* also featured numerous interviews with guests who affirmed that global warming is happening. One of these was former US vice president Al Gore, who visited both Stewart's program (November 4, 2009) and Colbert's (September 13, 2011) to discuss the evidence for human-caused climate change. Other guest affirming the existence of climate change included Secretary of Energy Steven Chu, who gives Stewart an overview of President Barack Obama's "cap-and-trade" proposal for limiting greenhouse gas emissions (July 21, 2009), and professional surfer Laird Hamilton, who tells Colbert that "global warming can create bigger storms, and has, and is going to" (December 15, 2010). The hosts give these guests friendly receptions. By contrast, Stewart takes a confrontational stance toward climate change skeptic Christopher Horner in a February 13, 2007 interview, challenging his arguments ("You're saying it's profit motive for all these scientists?") and mocking his book for its lack of data.

Colbert has kept up the heat on climate change skeptics as host of *The Late Show*. For example, a December 14, 2018 episode features a cartoon parody in which Frosty the Snowman scoffs at global warming: "Everything's fine," he says, even after he's melted. A February 22, 2019 episode uses another cartoon parody—this time of the children's television show *Captain Planet*—to mock President Donald Trump's appointment of global warming skeptics to a special committee studying climate change:

VOICE-OVER: Our world is in peril! President Trump sends five special rings to form his climate committee. Chip, Eric Trump's college drinking buddy. A woman who sort of looks like Ivanka. Ira, a guy Trump met in the steam room at Mar-a-Lago. Kid Rock. And William Happer, an unsightly climate change denier.
FAKE HAPPER: If climate change is real, how come it snows? I'm gonna go punch a dolphin.
VOICE-OVER: Together, they summon the power of willful ignorance!

Nor is Colbert the only late-night television host delivering such messages. For example, Samantha Bee devotes a November 15, 2017 segment of her program, *Full Frontal*, to how rising sea levels from climate change are gradually inundating Tangier Island, Virginia. In this segment, correspondent Allana Harkin interviews a marine biologist who's wearing sunglasses, a straw hat, and pastel beachwear that shows off his large biceps:

HARKIN: David Schulte is a scientist with the Army Corp of Engineers and world's strongest Jimmy Buffet fan. He studied the island for fifteen years.

SCHULTE: The majority of [residents] simply cannot accept that climate change is a real problem out here. They attribute it all to erosion.

In the same vein, Trevor Noah points out in an October 16, 2018 episode of *The Daily Show* that climate change may threaten the world's beer supply: "If you tell Americans that the Marshall Islands will be underwater in ten years, no one cares," he jokes, "but tell them Corona will cost more and now they're marching in the streets!"

With these sorts of messages in mind, we used our fall 2016 survey of the American public to test whether watching late-night television is linked to believing in human-caused climate change.[14] At first glance, the answer seems to be yes (Figure 6.1). Among respondents who said they get information about

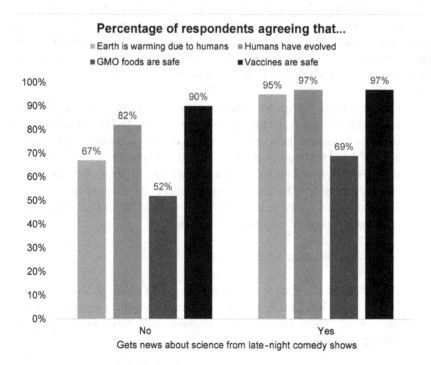

FIGURE 6.1 Science beliefs, by late-night comedy show watching (Cooperative Congressional Election Survey, 2016)

science from late-night comedy shows such as *The Late Show* or *Last Week Tonight*, fully 95% believed the earth is getting warmer because of human activity (Figure 6.1). Meanwhile, the figure was almost 30 points lower (67%) for non-late-night viewers. This gap was even bigger than the ones we found for CNN viewers versus non-viewers and MSNBC viewers versus non-viewers.[15]

Yet the link between watching late-night comedy shows and believing in human-caused climate change could reflect who tunes into such programs. People with varying degrees of interest in science watch them, but their audiences tend to be relatively liberal, Democratic, and non-religious—three characteristics that also go hand in hand with believing in climate change. For a different angle on late-night comedy effects, let's see what happened when we showed people specific segments about global warming and then measured their perceptions afterward.

Two Styles of Climate Satire

Here, we should point out that late-night hosts use two different types of satirical humor: sarcasm and irony. The first of these offers clear signals through tone of voice and gestures (eye rolls, air quotes, and so on) about what the comedian truly means to say.[16] During his tenure at *The Daily Show*, Jon Stewart often took this approach in discussing climate change. For example, he uses sarcasm in his October 26, 2011 description of a new global warming study:

> If only an impartial arbiter could come in—remove the warming debate's political implications and just examine the science. And if only that person could be funded in large part by two titans of a seldom-heard constituency of the global warming debate: the oil industry. Yes, Richard Muller, the Berkeley physics professor who took on the challenge of reexamining climate data had, as his biggest private funder, the Koch brothers ... Oil billionaires and Tea Party heartthrobs Charles and David Koch. So you see where this "research" is going.

Stewart goes on to describe Muller's results:

> Whoa, global warming is real. Did not see that coming. Yes, the study, funded by the Koch brothers, confirms that the original research was actually correct. The earth is getting warmer—or, judging by this graphic, getting more embarrassed.

Another style of satire revolves around irony, in which the comedian says one thing with a straight face while expecting the audience to draw the opposite conclusion.[17] In his character of a conservative talk show host on *The Colbert*

Report, Stephen Colbert often took this deadpan approach to affirming climate change and mocking skeptics. For example, he begins a January 28, 2013 mono-logue by feigning disbelief in global warming:

> Last week, President Obama cynically used the inaugural address to push his radical pro-survival agenda … Folks, I didn't think this part of his speech would get any traction because there's no national consensus on climate change. It's like if J.F.K. announced the Apollo program, but half the country denied the moon exists.

Colbert then cites Muller's study:

> But … even Koch brothers-funded climate change skeptic and hairbrush denier Richard Muller has done a 180, now stating, "Global warming is real, and humans are almost entirely the cause." Now the only thing receding faster than the glaciers is Dr. Muller's funding.

So, did viewers get these jokes, and were they swayed by Stewart's and Colbert's satire? In studying the effects of late-night comedy, it's important to remember two points. The first is that humor can prompt people to pay more attention to messages while also disarming their resistance to information that clashes with their pre-existing beliefs. As Dannagal Young argues in her book *Irony and Outrage,* the playfulness and incongruity of late-night satire motivate audience members to reinterpret information and fill in missing blanks so they can under-stand the humor. In doing so, viewers may help persuade themselves to accept the humor's underlying message instead of forming counterarguments against it.[18] With its built-in contradictions and complex meanings, the sort of irony that Colbert uses on *The Colbert Report* can be especially powerful in prompting viewers to engage with topics such as science.

The second key to understanding the effects of late-night satire is that audi-ence members may interpret humor in different ways. This is particularly likely when the jokes are ambiguous, as ironic humor often is. Indeed, a study by Heather LaMarre and her colleagues found that conservative viewers were more likely than liberal ones to think that Colbert really meant what he said when he mocked liberals and praised conservatives on *The Colbert Report.*[19] By the same logic, audience members could view Colbert's climate change humor through the prism of their own political beliefs.

In 2013, one of us (Paul) and Jessica McKnight conducted an experiment to capture the effects of late-night satire on climate perceptions.[20] We randomly divided 424 college students into three groups (students may not be typical of all viewers, but they're a key demographic for late-night comedy shows). One group viewed the *Daily Show* clip in which Stewart talks about Muller's study. Another group saw the *Colbert Report* clip in which Colbert discusses the same study. The third group—the control condition—watched a video on an unrelated topic and served as the baseline for testing the effects of late-night humor.

Most of the viewers in the study understood Stewart's sarcastic humor about climate change. Two-thirds of the participants who watched the *Daily Show* clip (67%) correctly identified him as believing that global warming is happening, while 23% said his stance was unclear and only 10% said he doesn't think global warming is happening. Furthermore, viewers interpreted Stewart's message in largely similar ways regardless of their own political views: 73% of liberals, 67% of moderates, and even 61% of conservatives saw him as believing in climate change (Figure 6.2).

The picture was different for Colbert's ironic humor. Only half (52%) of the viewers in the study saw him as believing that global warming is happening. A third (32%) said it wasn't clear what he thought, and another 16% said he believed that global warming isn't happening. Moreover, some viewers seem to

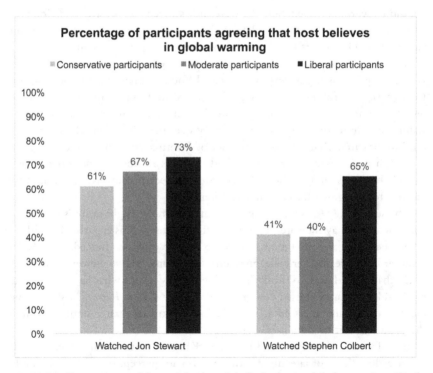

FIGURE 6.2 Perceptions of late-night hosts' beliefs about global warming (*Daily Show/Colbert Report* experiment, 2013)

have heard what they wanted to hear based on their own political beliefs: only 41% of conservative viewers perceived Colbert as believing in global warming, compared to 65% of liberal viewers (Figure 6.2). This pattern highlights one risk in using ironic humor: not every audience member will interpret the same joke in the same way.

Even so, Stewart's relatively overt commentary and Colbert's more dead-pan jokes yielded similar overall effects. Fully 82% of the participants who watched the *Daily Show* clip and 79% of the ones who watched the *Colbert Report* clip said that global warming is happening, compared to only 68% of those in the baseline group. In short, both sarcastic and ironic humor from these programs influenced viewers to accept the scientific consensus on climate change.[21]

"A Statistically Representative Climate Debate"

Late-night shows could also shape perceptions of climate science itself. Take *Last Week Tonight*, which has addressed global warming in several episodes—including in its May 11, 2014 one, which takes a clever approach to communicating the scientific consensus on the topic. John Oliver begins the episode with the sort of broad humor that might draw in viewers seeking to be entertained, not necessarily informed:

> The Earth: You may know it as that blue thing that Bruce Willis is always trying to save, or from its famous collaboration with wind and fire, or just simply as that place where George Clooney lives. Anyway, the Earth had some genuinely bad news this week.

From there, Oliver turns to a White House report concluding that global warming threatens every part of the United States and is "affecting us now." "Smart move, Obama," he says. "That is a key shift in how to talk about climate change because we've all proven that we cannot be trusted with the future tense." After pointing out that "one in four Americans is skeptical about climate change," Oliver describes the scientific evidence on global warming:

> There is a mountain of research on this topic. Global temperatures are rising, heat waves are becoming more common, sea surface temperatures are also rising, glaciers are melting, and, of course, no climate report is complete without the obligatory photo of a polar bear balancing on a piece of ice … A survey of thousands of scientific papers that took a position on climate change found that 97% endorsed the position that humans are causing global warming.

Next, Oliver turns to criticizing television news coverage of climate change, echoing media researchers' concerns about "balance as bias."[22] "I think I know why people still think this issue is open to debate," he says, "because on TV it is, and it's always one person for and one person against, and it's usually the same person for." Here, the segment cuts to a montage of clips from cable television news programs, each of which features a debate between a climate change skeptic and "Bill Nye the Science Guy." "Yeah, that's right," Oliver says, "more often than not it's Bill Nye the Science Guy versus some dude, and when you look at the screen, it's 50-50, which is inherently misleading."

At this point, the host introduces the concept of a "statistically representative climate change debate" to satirize balanced media coverage and affirm the scientific consensus, arguing that "if there has to be a debate about the reality of climate change—and there doesn't—then there is only one mathematically fair way to do it." The segment cuts to Oliver seated at a table, flanked by Bill Nye and another person playing the role of a climate change skeptic:

OLIVER: Good evening. Joining me tonight, a climate change denier, and, naturally, Bill Nye [the] Science Guy.
NYE: John, humans are causing climate change, no question …
OLIVER: Wait, wait, before we begin, in the interest of mathematical balance I'm going to bring out two people who agree with you, climate skeptic, and Bill Nye, I'm also going to bring out 96 other scientists. It's a little unwieldy, but it's the only way to actually have a representative discussion.

As Oliver speaks, the stage fills with scientists wearing lab coats. The host then addresses the skeptic, now seated with two other people:

OLIVER: Climate change skeptic, please make a case against climate change.
SKEPTIC: Well, I just don't think all of the science is in yet.
OLIVER: And what is the overwhelming view of the entire scientific community?

All the scientists start talking at once. Oliver asks the skeptic, "Any response to that?" The skeptic attempts to reply, but the host shouts, "I can't hear you over the weight of scientific evidence!" As the segment concludes, he continues to shout, saying, "This whole debate should not have happened. I apologize to everyone at home. My thanks to Bill Nye and the overwhelming scientific consensus."

By emphasizing the agreement among scientists that humans are causing climate change, Oliver's satire follows the advice of science communication researchers who advocate "consensus messaging" on the topic. In particular, these scholars argue that getting people to recognize the scientific consensus

on climate change is a key step toward persuading them that humans are causing climate change—and, ultimately, for persuading people to take action
on the issue.[23] Research on consensus messaging has typically looked at
non-humorous messages, but could a consensus message with a satirical spin—
specifically, Oliver's "statistically representative climate debate"—have a
similar effect?

In 2014, one of us (Paul again) and Jessica McKnight recruited a new sample of 288 college students and randomly split them into two groups.[24] The
first group viewed the *Last Week Tonight* segment; meanwhile, the second group
didn't watch any video about climate change. Compared to participants in
the baseline group, the ones who watched the "statistically representative climate debate" were more likely to say that most *scientists* think human-caused
global warming is happening (by more than half a point on a nine-point scale;
Figure 6.3). Furthermore, accepting the scientific consensus on the issue led
these participants to believe in human-caused climate change, too—just as the
advocates of consensus messaging would predict.

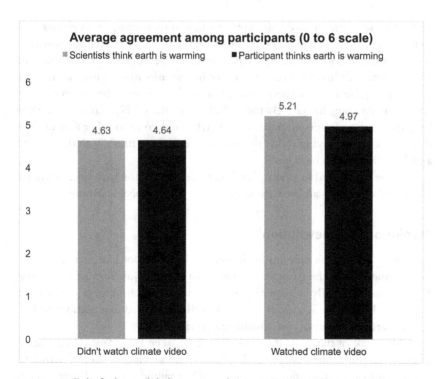

FIGURE 6.3 Beliefs about global warming, by *Last Week Tonight* watching condition
(*Last Week Tonight* experiment, 2014)

The study also yielded additional signs of how late-night humor can influence the sorts of people who don't usually engage with science and are less likely to hear what scientists say about key issues. In keeping with the "gateway hypothesis" proposed by Feldman and her colleagues, the impact of Oliver's global warming humor was especially strong among audience members with low interest in science. Specifically, watching the clip bridged the gap between these viewers and high-interest viewers in recognizing the scientific consensus on climate change.[25]

Five years later, Bill Nye made another appearance on *Last Week Tonight* to discuss climate change and, in the process, satirize late-night television's fondness for illustrating scientific principles through funny demonstrations. This time, Oliver asks Nye to explain "the complicated logic behind carbon pricing":

NYE [DONS SAFETY GLASSES AND POINTS TO AN EASEL]: When something costs more, people buy less of it. Safety glasses off. That's it.

OLIVER: Honestly, I was expecting something a bit more fun and visual than that, Bill. Could you maybe explain the long-term impact of carbon-pricing, but with a cool stunt to jazz up what you're saying? Go on…

Nye complies with an in-depth explanation, and then adds, "And because, for some reason, John, you're a 42-year-old man who needs his attention sustained with tricks, here's some fucking Mentos and a bottle of diet Coke." At the end of the segment, Oliver brings on Nye again and asks him to use another one of his "enjoyable light-hearted demonstrations" to drive home the urgency of addressing climate change. "By the end of this century," Nye says, "if emission keep rising, the average temperature of Earth could go up another four to eight degrees. What I'm saying is that the planet is on fucking fire." With that, he sets a globe ablaze with a blowtorch.

Judging by our earlier results, Nye's fiery demonstration may have swayed at least some viewers to accept what scientists say about global warming.

"Evolution, Schmevolution"

Late-night television's take on evolution has looked a lot like its take on climate change. Over the past two decades, hosts such as Jon Stewart and Jimmy Kimmel have typically sided with the scientific consensus in favor of evolution while mocking those who tout alternative religious-based explanations such as young-Earth creationism and "intelligent design" (ID).

At the peak of the "evolution wars" in 2005, *The Daily Show* aired a series of episodes, under the banner of "Evolution Schmevolution Week," satirizing anti-evolution activists' efforts to promote ID in public schools. Whereas traditional news outlets often presented "balanced" coverage of the issue—thereby abetting ID proponents' call to "teach the controversy"—Stewart and his team emphasized the scientific consensus.[26] In one episode (from September 12, 2005), the host describes evolution as being "widely accepted by nearly all scientists" and calls ID "crazy-claim magic fun hour." Correspondent Ed Helms then pretends to mistake contemporary evolution skeptics for Scopes Monkey Trial re-enactors.

Another episode (from September 14, 2005) uses a panel discussion to mock not only ID but also the notion of "teaching the controversy." Stewart invites three panelists to discuss the origins of life: a scientist who explains evolution, an ID proponent, and an advocate of "metaphysical creation theory" who argues that "creation is like a ball of energy … … it's part of psychogeometry … it's virtual reality to the projected eye." When she finishes, Stewart asks the ID proponent, "Why shouldn't *that* be taught in schools?" By doing so, he challenges the idea that ID deserves equal time with evolution in both public schools and news coverage.

Since 2005, the battle over evolution has faded somewhat from the public eye—and from satirical comedy television, as well.[27] Still, late-night programs have occasionally featured jokes about the topic, particularly at the expense of anti-evolution activists. Consider the following exchange from Aasif Mandvi's 2011 "Science: What's It Up To?" *Daily Show* segment:

NIKPOUR: It's very confusing for a child to only be taught evolution—to go home to a household where their parents say, "Wait a minute, you know, God created the Earth."
MANDVI [DEADPAN]: What is the point of teaching children facts if it's just going to confuse them?

Similarly, the May 7, 2019 episode of *Jimmy Kimmel Live!* features a public service announcement, narrated by actor George Clooney, for a mock organization named United to Defeat Untruthful Misinformation and Support Science (UDUMASS). Clooney tells viewers that a $200 contribution to UDUMASS (try saying it out loud) will help "teach ten dip—knuckle-draggers that dinosaurs existed, but not at the same time as people."

Have late-night jokes swayed the public on evolution? Our fall 2016 survey found a 15-percentage point gap between late-night viewers and non-viewers on the question of whether humans and other living things have evolved over time: 97% versus 82% (Figure 6.1). Paralleling our findings for traditional news sources, the gap between late-night viewers and non-viewers was narrower on evolution than on climate change. [28] This gap narrowed even further when we accounted for other factors—such as demographics, political ideology, and religiosity—that also shape perceptions of evolution. Still, the influence of late-night television may have been stronger during the peak of the "evolution wars" in the early 2000s. If the results from our climate change experiments are any guide, sarcastic jokes, ironic humor, and mock debates about evolution could all influence viewers' beliefs about the origins of life.

"Be Like Ethan, Kids"

Just as late-night comedy shows have mocked anti-science voices on climate change and evolution, so, too, have they satirized the anti-vaccine movement. For example, a series of measles outbreaks in 2014—including one at Disneyland, California—prompted both Larry Wilmore and Jimmy Kimmel to address the issue. Wilmore's monologue from the January 27, 2015 episode of *The Nightly Show* mocks Jenny McCarthy for saying that she learned about a supposed vaccine-autism link through the "University of Google" ("I don't think it's a good idea to take medical advice from a non-doctor ... a lot of my old friends disagreed, and I miss them every day."), while the February 27, 2015 episode of *Jimmy Kimmel Live!* features a clip of frustrated doctors telling viewers things such as, "You remember that time you got polio? No, you don't, because your parents got you f—ing vaccinated."

Late night's advocacy for vaccines continued as the number of measles cases in the United States grew. As a case in point, the June 25, 2017 episode of *Last Week Tonight* devotes almost half an hour to the topic. "Much of the fear surrounding vaccines stems from their supposed link to autism," John Oliver explains. "Now, that is a theory that gained traction in the late 90s thanks to a study published in the medical journal *The Lancet* suggesting a link between the MMR vaccine and autism. The study was of just twelve children and it was by this guy: Andrew Wakefield." After explaining that Wakefield's medical title was revoked for fraud and unethical behavior—"He's basically the Lance Armstrong of doctors"—Oliver addresses some of the concerns behind vaccine hesitancy:

> I kind of get the insistence that there must be a link. The age children are supposed to get the MMR vaccine happens to be the same age that diagnosable symptoms of autism can begin to appear. But correlation is not causation. That is what scientific studies are for, and ... they are really clear: that link is not there.

He concludes the segment by emphasizing the enormous public health benefits of vaccines and describing his own experience as a parent: "If this helps at all, I have a son. He is nineteen months old. He was born prematurely following a very difficult pregnancy, and ... I still worry about his health a lot, but we are vaccinating him fully on schedule."

Similarly, the April 3, 2019 episode of *Full Frontal* highlights the importance of vaccines and satirizes "anti-vaxxers." Like Oliver, Samantha Bee emphasizes that Wakefield's MMR-autism study was "a false study by an unlicensed doctor." She also tells viewers: "Don't treat autistic people like their existence is worse than a pandemic." Bee then interviews Ethan Lindenberger, a teenager "who got himself vaccinated when his parents would not," and enlists him in an "anti-anti-vaxxer" skit based on the television show *Riverdale*:

BEE [PLAYING A HIGH SCHOOL CHEERLEADER]: Oh my God, Archie, are you hurt?
ARCHIE: The opposite, actually. I got vaccinated.
BEE: Vaccinated? I don't do that stuff. My parents don't approve. [Coughs blood into a handkerchief]
ARCHIE: I looked online and realized I don't want to die from tuberculosis. I would rather die in a more normal way: by a cult.
BEE: Wow, you make vaccinations sound so cool. Maybe it is bad to get diseases from the Middle Ages.

Bee ends the segment by saying, "Be like Ethan, kids!"

Almost every late-night viewer we polled in our fall 2016 survey—97% of them—agreed that vaccines are safe for healthy children (Figure 6.1). The figure was also high among non-viewers, but not quite as high, at 90%. As Oliver points out, correlation is not always causation; in this case, however, we found a clear gap between late-night viewers and non-viewers even when we accounted for other factors that could shape perceptions of vaccines, from education to news media use. By way of comparison, this gap was as wide—and as clear—as the one between science magazine readers and non-readers on the same topic.[29]

For some vaccine-hesitant viewers, then, humorous messages from the likes of Oliver and Bee may make a real difference in reinforcing the scientific consensus that vaccines for resurgent diseases such as measles are safe and beneficial. Likewise, late-night shows could help strengthen trust in vaccines for new diseases such as COVID-19 by providing platforms to pro-vaccine spokespeople (as Trevor Noah did in a November 20, 2020 interview of Bill Gates) and weaving pro-vaccine messages into their jokes (as Stephen Colbert did in a December 3, 2020 segment where he invited former presidents to get vaccinated on his show).

"What's a GMO?"

On October 9, 2014, Jimmy Kimmel tackled yet another science-related topic: genetically modified foods. "Critics of genetically modified organisms, or GMOs, claim that they pose health risks," he explains during his monologue. "It's a complicated issue, though, because there isn't a lot of evidence to back that up. Still, some people are dead set against them. There's been legislation designed to limit GMOs, you see documentaries and angry Facebook posts about it." The host then asks, "How many of you do not want GMOs in your body?" When members of the studio audience clap and shout, Kimmel gets to his point:

> As I usually do when people take a stance on a complicated issue, I wondered: "How many people who are against GMOs really know what they are?" So, we sent a crew to one of our local farmer's markets to ask people why they avoid GMOs and, more specifically, what the letters GMO stand for.

The first interviewee says he avoids GMOs because "there's just a vibration" with them. He correctly identifies the "genetically modified" part of the abbreviation, then stumbles on the O. A second interviewee cites "just the effects, I guess, on myself" as her reason for avoiding GMOs but can't explain what any of the letters stand for. Other interviewees featured in the segment likewise struggle to say anything coherent about the topic. The comedy of the segment comes from watching people flail as they try to explain themselves, but the underlying message of Kimmel's humor is that many GMO skeptics are hazy on the science behind agricultural biotechnology.

The Daily Show presents its own satirical look at GMO critics in its April 22, 2015 episode. Jon Stewart introduces the topic by saying, "If you spend a lot of time looking at food packaging—and I do, because you know when you're high you want to know what you're eating—you might think the three most terrifying letters in the English language are G-M-O. But W-H-Y?" From there, Aasif Mandvi investigates the US Food and Drug Administration's decision to approve the genetically modified Simplot potato. First, he interviews GMO critic Jeffrey Smith, who says that "genetically modified crops may be one of the most dangerous introductions of an additive to our food supply in our history." The correspondent then interviews a potato geneticist who debunks Smith's claims, explaining that "genetically modified foods crops don't alter human DNA." Mandvi concludes that the "Simplot's devil tater is actually an angel tot." He also points out that Jeffrey Smith is not a scientist and cites the Pew Research Center's finding that "88% of scientists believe GMOs are safe."

Whereas Jimmy Kimmel and *The Daily Show* have satirized GMO critics, other late-night television shows have occasionally given airtime to anti-GMO views. For example, the November 6, 2013 episode of *The Colbert Report* mocks arguments against labeling GMO foods, with the host ironically proclaiming that "it is none of our business what we are putting in our mouths." Likewise, the May 27, 2016 episode of Colbert's *The Late Show* gives musician Neil Young a platform for criticizing GMOs. The host mentions that one study found no nutritional difference between GMO and non-GMO diets, but he allows Young to reply, "That must be a Monsanto study." Young then debates an intern dressed as a genetically modified ear of corn who becomes so angry he explodes into popcorn.

On the whole, however, fans of satirical comedy tend to agree with Mandvi and the scientific community on GMOs. Around two-thirds (69%) of the late-night viewers we polled in 2016 saw GMO foods as safe; meanwhile, only around half (52%) of non-viewers agreed (Figure 6.1). On this issue, the gap between late-night viewers and non-viewers was around the same size as the ones between science magazine readers versus non-readers and newspaper readers versus non-readers.[30] Even after we controlled for other factors, including traditional media consumption, late-night viewers still stood out in terms of how much faith they placed in genetically modified foods. Here, as with vaccines, satirical humor may help reinforce acceptance of the scientific consensus by drawing in a broader audience and disarming resistance among viewers inclined to reject more conventional messages.

Looking beyond GMO foods, late-night comedy could also help engage viewers with more recent developments in biotechnology. Take *Last Week Tonight's* July 2, 2018 segment about CRISPR, a relatively inexpensive and easy-to-use gene-editing technology that has received growing media attention.[31] "We need to figure out how to balance the risks and potential rewards of gene editing," Oliver says. "Which is going to be tricky because everything that's being done tends to get mixed together [in media coverage]—meticulous professional scientists with freewheeling biohackers ... practical applications with wild theories, best case scenarios like ending malaria with catastrophic prophecies of 30-foot wolves." Throughout the segment, he satirizes implausible "runaway science" frames for CRISPR while encouraging audience members to weigh practical and ethical questions surrounding the technology.

Satire and Critical Engagement with Science

Late-night comedy shows occasionally stereotype scientists as ivory tower weirdos or promote questionable scientific claims. More often, however, these programs portray science as exciting, reinforce what scientists say about important issues, and debunk anti-science voices. As with the cases we've already explored, late-night coverage of COVID-19 illustrates these patterns. Over the course of 2020, *Last Week Tonight* ran 11 different segments on the virus, each of which was around 20 minutes long. Collectively, they use humor to explain the science of the virus, echo public health recommendations, and counter misinformation about the pandemic. For example, the program's July 20, 2020 episode features an in-depth rebuttal of conspiracy theories about COVID-19 such as the ones featured in the spurious documentary *Plandemic*. "The issue isn't just that the film misrepresents [conspiracy theorist Judy] Mikovitz's backstory," Oliver says, "it's that in doing so, they lend her an air of credibility when they allow her to make unchallenged batshit medical claims."

Many viewers are in on the joke, too. On issues such as climate change, vaccines, and GMOs, there's a symbiotic relationship between laughing about science and learning about it. Watching late-night comedy programs can foster interest in science as well as acceptance of consensus scientific viewpoints on important issues. Moreover, these shows seem to be particularly effective at swaying Americans who are less informed about and interested in science. By using satirical humor to reach such viewers, late-night comedians can speak to part of the "missing audience" for traditional science communication.

At its best, late-night satire not only promotes scientific engagement and awareness but also encourages critical thought about scientific messages. For example, John Oliver and Bill Nye's "statistically representative climate debate" challenges the "he said, she said" approach that traditional news outlets have sometimes taken in covering global warming. Likewise, Jon Stewart's three-way debate over evolution and ID skewers the notion of "teaching the controversy." Both segments call attention to how seemingly objective reporting can produce distorted coverage of what the science says.

The May 8, 2016 episode of *Last Week Tonight* provides a striking illustration of how humor can prompt viewers to engage in active deliberation about science itself along with news coverage of it. Unlike most media messages about science, this segment highlights—and satirizes—the social context in which scientific research occurs and the limitations of common scientific practices. Oliver begins by describing several common flaws in scientific studies covered by the news media. These include the bias toward publishing positive results ("Nobody is publishing a study that says, 'Nothing up with açaí berries'"), the potential misuses of statistics ("P-hacking … basically means collecting lots of variables and then playing with your data until you find something that counts as statistically significant but is probably meaningless"), and the absence of incentives for replication studies ("There's no Nobel Prize for fact-checking").

Oliver then discusses how news stories about scientific studies often selectively cover results ("In science, you don't just get to cherry-pick the parts that justify what you were going to do anyway") and sensationalize complex findings ("Too often, a small study with nuanced, tentative findings gets blown out of all proportion when it's presented to us, the lay public"). He spends much of the segment satirizing science coverage on television, particularly on morning news shows *Today* and *Good Morning America*, but he doesn't spare social media, either. "When studies aren't all over TV," he says, "they're blanketing your Facebook feed with alerts like 'Study finds liberals are better than conservatives at smizing,' 'Your cat might be thinking about killing you,' and 'scientific study shows that bears engage in fellatio'" (all real examples of headlines). So, let's take a cue from *Last Week Tonight* and look at messages about science on social media platforms such as Facebook, Twitter, Instagram, and YouTube—which, after all, is where many people (us included) often encounter Oliver's own brand of science communication.

Notes

1 Mooney, Chris, and Sheril Kirshenbaum, *Unscientific America: How scientific illiteracy threatens our future*, Basic Books, 2009, 37.
2 Mooney and Kirshenbaum, *Unscientific America*.
3 "Late night ratings, May 6–10, 2019: *Late Show* slips," *TV by the Numbers*, May 30, 2019, https://tvbythenumbers.zap2it.com/weekly-ratings/late-night-ratings-may-6-10-2019-late-show-slips.
4 See Chapter 5.
5 Guenther, Lars, "Science journalism," in *Oxford research encyclopedia of communication*, Oxford University Press, 2019, https://oxfordre.com/communication/communication/view/10.1093/acrefore/9780190228613.001.0001/acrefore-9780190228613-e-901.
6 Project for Excellence in Journalism, "Journalism, satire or just laughs? *The Daily Show* with Jon Stewart, examined," Pew Research Center, May 8, 2008, www.journalism.org/node/10953.
7 Feldman, Lauren, Anthony Leiserowitz, and Edward Maibach, "The science of satire: *The Daily Show* and *The Colbert Report* as sources of public attention to science and the environment," in *The Stewart/Colbert effect: Essays on the real impacts of fake news*, ed. Amarnath Amarasingam, Jefferson, McFarland and Company, 2011.
8 Feldman et al., "The science of satire."
9 Feldman, Lauren, "Assumptions about science in satirical news and late-night comedy," in *The Oxford handbook of the science of science communication*, ed. Kathleen Hall Jamieson, Dan Kahan, and Dietram A. Scheufele, Oxford University Press, 2017, 321–332.
10 Baum, Matthew, *Soft news goes to war: Public opinion and American foreign policy in the new media age*, Princeton University Press, 2003; Feldman et al., "The science of satire."
11 Colbert mocked the former in the September 12, 2018 episode of *The Late Show* and the latter in the program's March 28, 2019 episode.
12 Brewer, Paul R., Dannagal G. Young, Jennifer L. Lambe, Lindsay H. Hoffman, and Justin Collier, "'Seize your moment, my lovely trolls': News, satire, and public opinion about net neutrality," *International Journal of Communication* 12 (2018); Moy, Patricia, Michael A. Xenos, and Verena K. Hess, "Priming effects of late-night comedy," *International Journal of Public Opinion Research* 18, no. 2 (2006): 198–210; Young, Dannagal Goldthwaite, "Late-night comedy in election 2000: Its influence on candidate trait ratings and the moderating effects of political knowledge and partisanship," *Journal of Broadcasting & Electronic Media* 48, no. 1 (2004): 1–22.
13 Feldman, Lauren, "Cloudy with a chance of heat balls: The portrayal of global warming on the daily show and the Colbert report," *International Journal of Communication* 7 (2013).
14 See the Appendix for survey details.
15 See Chapter 5.
16 LaMarre, Heather L., Kristen D. Landreville, and Michael A. Beam, "The irony of satire: Political ideology and the motivation to see what you want to see in *The Colbert Report*," *International Journal of Press/Politics* 14, no. 2 (2009): 212–231.
17 LaMarre et al., "The irony of satire."
18 Young, Dannagal Goldthwaite, *Irony and outrage: The polarized landscape of rage, fear, and laughter in the United States*, Oxford University Press, 2019.
19 LaMarre et al., "The irony of satire."
20 Brewer, Paul R., and Jessica McKnight, "Climate as comedy: The effects of satirical television news on climate change perceptions," *Science Communication* 37, no. 5 (2015): 635–657.
21 See also Becker, Amy, and Ashley A. Anderson, "Using humor to engage the public on climate change: The effect of exposure to one-sided vs. two-sided satire on message

discounting, elaboration and counterarguing," *Journal of Science Communication* 18, no. 4 (2019): A07; Skurka, Chris, Jeff Niederdeppe, and Robin Nabi, "Kimmel on climate: Disentangling the emotional ingredients of a satirical monologue," *Science Communication* 41, no. 4 (2019): 394–421.

22 Boykoff, Maxwell T., and Jules M. Boykoff, "Balance as bias: Global warming and the US prestige press," *Global environmental change* 14, no. 2 (2004): 125–136.

23 Cook, John, and Peter Jacobs, "Scientists are from Mars, laypeople are from Venus: An evidence-based rationale for communicating the consensus on climate," *Reports of the National Center for Science Education* 34, no. 6 (2014); van der Linden, Sander L., Anthony A. Leiserowitz, Geoffrey D. Feinberg, and Edward W. Maibach, "The scientific consensus on climate change as a gateway belief: Experimental evidence," *PloS one* 10, no. 2 (2015): e0118489.

24 Brewer, Paul R., and Jessica McKnight, "'A statistically representative climate change debate': Satirical television news, scientific consensus, and public perceptions of global warming," *Atlantic Journal of Communication* 25, no. 3 (2017): 166–180.

25 See also Anderson, Ashley A., and Amy B. Becker, "Not just funny after all: Sarcasm as a catalyst for public engagement with climate change," *Science Communication* 40, no. 4 (2018): 524–540.

26 See Chapter 5.

27 See Chapter 5.

28 See Chapter 5.

29 See Chapter 5.

30 See Chapter 5.

31 Annenberg Public Policy Center of the University of Pennsylvania, "Media framing of news stories about the ethics, benefits, and risks of CRISPR," January 2, 2021, https:// www.annenbergpublicpolicycenter.org/feature/science-media-monitor-report-2.

7

SOCIAL MEDIA SCIENCE

We don't just love science, we FUCKING LOVE IT. Science is mindblowingly, head fuckingly amazing. People need to know this.

—*Elise Andrew* (I Fucking Love Science, September 5, 2012)

Recently, I received a question for an "Ask Emily" episode along the lines of whether or not I had personally experienced sexism in the field. And I kind of shrugged it off ... The more I thought about it, though, along with another question of is there any part of my job that I don't look forward to, I would have to say it would be the frustratingly negative and sexist comments I have to sift through in my various inboxes on a daily basis.

—*Emily Graslie* ("Where My Ladies At?" *The Brain Scoop*, November 27, 2013)

During her final year as a biology student at Sheffield University, Elise Andrew created a new Facebook page titled I Fucking Love Science and posted a picture of two hands holding the Earth. The next day, she posted a quotation from author Isaac Asimov that encapsulated the philosophy of her site: "The most exciting phrase to hear in science, the one that truly heralds new discoveries, is not 'Eureka!' but 'That's funny.'" Many more posts followed. Some featured photographs of spiders or planets; others shared cartoons or memes; still others criticized anti-science politicians or voiced support for NASA's programs and for other scientific endeavors.

Despite—or perhaps partly because of—the expletive in its name, I Fucking Love Science (or IFLS, for short) swiftly became a social media sensation. Within

DOI: 10.4324/9781003190721-7

a year of its March 9, 2012 launch, the page had received millions of "likes" on Facebook and won praise from prominent voices in the scientific community, including Bill Nye and Neil deGrasse Tyson.[1] Andrew also branched out to other social media platforms, such as Twitter. "I'm not necessarily trying to teach people on IFLS," she explained in an interview. "But I am trying to just pique their interest a bit and show them that it's not as exclusive. It's not this old boys' club with old men sitting around stroking their white beards, talking about how clever they are."[2]

Andrew maintained a low personal profile during her first year of running of the site, but in early 2013 she shared a photograph of herself on Twitter. Many social media users reacted by remarking on her gender—and, in some cases, her appearance. "Dude, you're a chick? WTF," wrote one user. "Wait ... you're a chick? And you're hot?! Lol," commented another.[3] "I don't usually go off on a social sciences tangent," Andrew posted on March 20, 2013, "but ... it's a sad day when a woman being funny and interested in science is considered newsworthy."

While some social media users responded to her posts with sexist comments, others criticized Andrew for emphasizing "clickbait" that sensationalized or oversimplified scientific discoveries.[4] None of this, however, dampened IFLS's popularity. By 2017, it had amassed 24 million followers on Facebook and was the third most popular science-themed site on the platform, eclipsed by only the pages for National Geographic and the Discovery Channel.[5]

The rise of IFLS illustrates the growing importance of social media platforms as sources of scientific information for the public. A 2017 Pew Research Center survey found that more than three-fourths (79%) of social media users had seen science-related posts—and that a quarter (26%) followed science-themed pages or accounts.[6] When we asked about specific social media platforms in our November 2016 survey, 28% of the respondents said they got news or information about science from Facebook and 24% said the same about YouTube.[7] One in ten respondents said they learned about science from Twitter (10%), and about half that many received news about the subject from Instagram (4%) or Reddit (4%).

Given how many Americans encounter scientific information on social media platforms, messages on these platforms may shape public perceptions of science. In part, such influence could reflect the same sorts of effects produced by television news and other "old media." For example, frames in social media messages could sway what audience members think about scientists and scientific research. Unlike more traditional forms of media, however, social media platforms give a wider range of users—including both scientists and laypeople—opportunities to create and share scientific content.[8] Another feature that sets social media platforms apart from traditional channels for science communication is the way they allow users to interact with one another through informal conversations and structured online communities.

As researchers such as Zeynep Tufekci have pointed out, each social media platform also offers its own distinct set of affordances—that is, "things that it allows and makes easy versus things that are not possible or difficult."⁹ Take Twitter and Instagram: both sites allow users to include hashtags that can increase a message's visibility, but Twitter's design emphasizes short, text-based messages whereas Instagram's design emphasizes image sharing. The different communication styles these platforms encourage could help determine how messages on them shape perceptions of science. Furthermore, the specific affordances that social media sites offer for user interaction and engagement may create multiple pathways for influence. For example, Reddit hosts forums about science with millions of members and features a system for upvoting or downvoting posts, while YouTube displays user comments along with the number of views, likes, and dislikes for each video hosted on the site. Such social cues from other site members could sway how Reddit and YouTube users form their impressions of science.¹⁰

To understand what sorts of scientific messages—and messengers—audience members see and engage with on social media platforms, we draw on big-picture data as well as a variety of case studies. We also look at how social media messages influence users' perceptions of science. For example, do the lighthearted posts featured on IFLS spark more positive views of scientists? And can scientists come across as more trustworthy and relatable by posting Instagram "selfies?"

In some instances, science communication on social media carries troubling implications for both the scientific community and broader society. As a case in point, social media can reflect—and reinforce—the gender disparities and sexism prevalent in science. Furthermore, social media sites often serve as channels for scientific misinformation on topics ranging from climate change to COVID-19. Yet social media messages can also help address these same problems. As we show, communicators such as Emily Graslie from the YouTube channel *The Brain Scoop* have used social media to challenge sexism in science. Similarly, a look at recent evidence reveals that both algorithm-generated links and user-written comments can help correct false messages about topics such as vaccines and GMOs.

In sorting out the complex and sometimes contradictory roles that social media messages play in shaping perceptions of science, we start with a familiar concept from research on traditional news media effects: framing. Specifically, we look at what frames appear on some of the most prominent science-themed Facebook pages—including Elise Andrew's page.

Framing Science on Facebook

In 2017, the Pew Research Center conducted a study of more than 130,000 posts on 30 popular science-related Facebook pages, included the ones for National Geographic (which had 44 million followers at the time), the Discovery Channel

(39 million), Animal Planet (20 million), and NASA (19 million), along with IFLS.[11] To see how these pages framed science, the researchers analyzed a random sample of posts from each page. The top frame, appearing in 29% of all posts, revolved around new scientific discoveries. For example, a December 8, 2016 post on National Geographic's page showed a recently unearthed dinosaur tail preserved in amber. Some pages heavily emphasized the "new discovery" frame, including IFLS, NASA, and National Geographic. By contrast, the Discovery Channel and Animal Planet highlighted this frame in only a handful of posts.

Another common frame was "news you can use" featuring tips and advice for audience members (21% of all posts). Whereas pages such as IFLS, NASA, and National Geographic seldom relied on this frame, other science-themed pages—such as Psychology Today and Health Digest—highlighted it in a majority of their posts. For example, a January 26, 2017 post from Health Digest touted the "Many Health Benefits of Babaco" (a fruit related to the papaya). The pages for two prominent health commentators, Dr. Mehmet Oz and raw food advocate David "Avocado" Wolfe, also featured many posts with "news you can use." Some of these posts included dubious scientific claims, with Wolfe—an anti-vaxxer and flat earth proponent—offering especially suspect advice.

Still other pages focused on frames promoting television shows or public appearances. More than half the posts on the Discovery Channel and Animal Planet's pages pitched programs on these networks, as when the former shared an image from the series *Deadliest Catch* with the caption, "Seeing good crab never gets old!" Similar types of self-advertisements were common on the Facebook pages for several prominent science communicators, including Bill Nye, Neil deGrasse Tyson, and Stephen Hawking. Overall, 16% of posts featured a promotional frame.

Building on Pew's findings, we conducted an experiment in May 2020 to test how frames from three of the top science-themed Facebook pages—IFLS, NASA, and National Geographic—shaped internet users' views of scientists and scientific research.[12] We looked at the most common frame on these pages, the "new discovery" frame, and two of the most common topics on them: astronomy and animal science. As part of a national online survey, we randomly assigned respondents to see different posts from the three pages. Some respondents viewed an IFLS post headlined, "NASA Share[s] Image of 'Unusual' Hole on Mars," along with an image of the hole and the caption "NOT NOW, ALIENS!" (the post was from March 15, 2000, in the midst of the COVID-19 pandemic). A second group of respondents saw a different post from IFLS featuring a photograph of a pink manta ray and the caption, "Animals with this unusual coloring are easy prey and generally don't make it to adulthood, but this male ray is doing just fine."[13] Another group viewed a National Geographic post about the same pink manta ray, and yet another group saw a NASA post about tectonic activity on Mars. Meanwhile, the control group didn't receive any Facebook post.

The relatively staid posts from NASA and National Geographic didn't influence respondents' views about scientists. Nor did these posts sway opinions about research on space or sea life. On the other hand, the quirkier posts from IFLS did shape views of scientists and science (Figure 7.1). Compared to respondents in the other groups, the ones who saw the IFLS posts about the hole on Mars or the pink manta ray were eight to nine percentage points more likely to agree that scientists work for the good of humanity. Similarly, the IFLS groups were five to six points likelier to agree that scientists care about helping the public understand new research. Seeing the IFLS post about the manta ray didn't significantly boost support for research on sea life, but the IFLS post about the hole on Mars did produce an eight-point bump in support for spending on space exploration.

In short, Elise Andrew's lighthearted approach to framing science through Facebook influenced audience members. By using humor to engage their

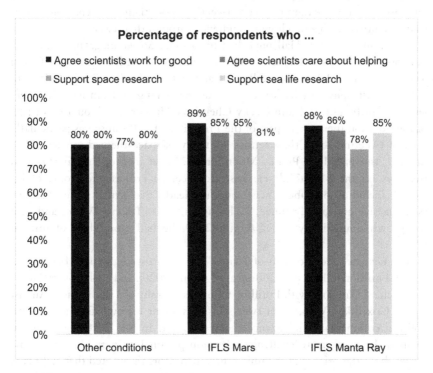

FIGURE 7.1 Perceptions of and opinions about science, by Facebook viewing condition (Center for Political Communication Survey, 2020)

attention and convey her own love of the subject, her page encouraged them to love—or at least like—science and scientists a little more, too. But what happens when scientists themselves use social media to communicate with broader audiences? To learn more about this, let's shift from Facebook to a platform that many scientists use to engage with one another and the public: Twitter.

Twitter Stars and Tweets about Stars

In 2014, genomics researcher Neil Hall published an article proposing a new measure of social media status among scientists: the "Kardashian index," named for celebrity Kim Kardashian.[14] Hall pointed out that some scientists on Twitter have more followers than scholarly citations (the currency of scientific influence), whereas other scientists have more citations than followers. He also suggested calculating the ratio of followers to citations to identify the "Kardashians" of science—that is, the scientists who are famous for being famous, rather than for their research. Though Hall was joking, his article prompted a serious discussion of how Twitter expands, and perhaps distorts, opportunities for scientists to communicate with their peers and the broader public. Within their own fields, scientists can enhance their impact by tweeting about their research.[15] Moreover, some scientists reach a much wider audience through their tweets.

A few months after Hall published his article, *Science* magazine named its "top 50 science stars of Twitter."[16] The first entry on their list was Neil deGrasse Tyson, who had 2.4 million followers at the time. Since joining Twitter in 2009, the astrophysicist has tweeted about topics ranging from new discoveries to scientific puns ("More Geeky Chemistry Humor ... If you are not part of the Solution then you're part of the Precipitate").[17] Tyson seems particularly fond of tweeting about the scientific accuracy of space-themed movies ("The film *Ad Astra* has Moon Pirates. Moon Pirates? What are they thinking? Buried treasures on the Moon?").[18] He has also received criticism for some of his tweets, including one about mass shooting deaths that struck some observers as insensitive and poorly timed.[19] Through all this, Tyson's Twitter audience has grown dramatically: as of August 2020, he had more than 14 million followers.

The next two "science stars of Twitter" on the list were physicist Brian Cox, with 1.4 million followers (3 million as of August 2020), and evolutionary biologist Richard Dawkins, with 1 million followers (just shy of 3 million as of August 2020). Cox, like Tyson, often tweets insights about his own field along with political commentary and science-themed humor. In July 2018, for example, he used the platform to challenge television personality Piers Morgan's comments about the origins of the universe. When Morgan tweeted that "[a]theists can never say what was there before the Big Bang," Cox replied, "If you mean the Hot Big Bang then there may have been a period of rapid expansion before known as inflation. This theory is able to account for observed features of the universe."[20]

Dawkins, meanwhile, is famous—or notorious—for his tweets about science and society. An avowed atheist, he frequently uses the site to express scathing views on religion in general and Islam in particular.[21] He has provoked controversy with tweets about topics such as rape and "political correctness," as well.[22] In February 2020, he sparked a new firestorm with a tweet about eugenics. "It's one thing to deplore eugenics on ideological, political, moral grounds," he wrote. "It's quite another to conclude that it wouldn't work in practice."[23]

The Twitter accounts of Tyson, Cox, and Dawkins illustrate a broader pattern: scientists use the platform to comment on a mix of topics, including subjects beyond their own fields. When Qing Ke and her colleagues analyzed tweets from more than 45,000 scientists in 2017, they found that scientists' tweets "blurr[ed] boundaries between the personal and the professional."[24] "It's about your coffee," one of the authors, Cassidy Sugimoto, told *Nature* reporter Sarah McQuate. "It's the news. It's politics. It's not necessarily scientific in nature."[25] However, some of the scientists in the sample did share links to scientific studies. As McQuate points out, Twitter also provides scientists with a platform for explaining research to non-scientists in an accessible way.

To find out what happens when scientists tweet about their fields to a broader audience, we included three additional groups in our May 2020 online survey. Each of these groups read a series of tweets from a high-profile astronomer about the giant star Betelgeuse. The first group saw tweets from Neil deGrasse Tyson, *Science* magazine's top "science star" of Twitter. The second group saw tweets from Phil Plait (#5 on the same list) and the third read tweets from Pamela Gay (ranked #33). Each set of tweets included a photograph of Betelgeuse along with an explanation of what makes the star so special: it's going to explode soon (in astronomical terms).

We thought the idea of a star "going boom," as Gay put it, was fascinating, but our respondents weren't as impressed. Their perceptions of scientists and space exploration didn't differ regardless of whether they read tweets about Betelgeuse from Tyson, Plait, or Gay. One lesson here is that not every social media message about science will influence audience members. Yet even if astronomers' tweets about a supernova didn't sway our participants, other sorts of messages from scientists on the platform—such as the more personal tweets described by Ke and her colleagues—could influence users. Given how IFLS's lighthearted post about Mars and manta rays boosted the image of science, maybe Tyson's tweets about movies or Gay's tweets about searching for kitschy rockets at Florida's Space Coast would do the same.

Messages from scientists on other social media platforms could also influence audience members, though the different affordances of these platforms can spur different sorts of messages. For example, Instagram's design encourages science communicators to take a more visual approach on the site—one that may help shape users' mental images of scientists.

Scientist Selfies and Instagram Influence

Many social media users probably associate Instagram with celebrities promoting themselves and "influencers" promoting products, but scientific agencies such as NASA and science media organizations such as National Geographic have built massive followings of their own on the photograph-sharing platform. Furthermore, thousands of individual scientists have used it to post pictures of their work along with "selfies" from their labs and field sites.[26]

NASA joined Instagram in 2013 with four photographs of the moon, including one of an "earthrise" taken by the Apollo 11 mission. The captions for the images publicized the upcoming launch of a new probe, the Lunar Atmosphere and Dust Explorer. Since then, NASA has posted thousands of images on Instagram. As of August 2020, its main account had more than 60 million followers, making it the 39th most followed account on the site.[27] Other accounts for the agency's programs and personnel have amassed hundreds of thousands of followers, as well.

Collectively, these Instagram accounts help promote NASA's endeavors—which, not coincidentally, depend on public funding. For example, the agency frequently posts photographs of planets, moons, and asteroids taken by its space probes, as well as shots of auroras, nebulae, and galaxies taken by its telescopes. NASA also uses the site to share images of its astronauts in dramatic settings such as space walks and more down-to-earth ones such as Halloween parties (though in space, so not *literally* down-to-earth). "Instagram's a very logical fit because of the visual nature and the imagery we ... have," explained Jason Townsend, NASA's deputy social media manager, in a 2017 interview. "We look to Instagram to showcase a photo each day."[28]

National Geographic has even more followers on the platform than NASA—around 137 million of them as of August 2020, placing it at #11 on the list of top accounts (ahead of both Taylor Swift at #13 and Kendall Jenner at #14).[29] National Geographic's most popular photographs on Instagram have included images of a diver with a tiger shark, a wolf in Yellowstone Park, sled dogs in Greenland, and a wildlife ranger with the last living male white rhinoceros.[30] All told, the organization's posts have received more than 4 billion "likes."[31]

No individual scientist comes close to matching NASA's or National Geographic's presence on Instagram, but thousands of researchers and science communicators have created accounts to share their work with followers.[32] Given that the site is especially popular among women, it's not surprising that many of these accounts belong to women.[33] For example, neuropsychologist Janelle Letzen (@the_sushi_scientist) uses photographs of sushi to illustrate topics such as the effects of serotonin and dopamine. Likewise, biologist Tagide deCarvalho (@nerd.candy) posts colorful photographs of microscopic organisms, and molecular geneticist Samantha Yammine (@science.sam) posts selfies with captions on subjects ranging from "the science of plant stem cells in skin care" to the biochemistry of "runner's high."

Despite the growing popularity of Instagram among scientists, not everyone in the profession takes a positive view of science communication on the site. In 2018, *Science* magazine published an opinion piece by engineering researcher Meghan Wright that criticized the use of Instagram by scientists, particularly women scientists. "By visibly contradicting stereotypes about female scientists, it is clear that they hope to inspire girls to pursue science and to encourage female scientists to showcase their femininity in our male-dominated work spaces ..." she argued. "But it disturbs me that these efforts are celebrated as ways to correct for the long held and deeply structured forms of discrimination and exclusion that female scientists face ... Time spent on Instagram is time away from research, and this affects women in science more than men."[34] Wright singled out Yammine's "Science Sam" account to illustrate her point.

Yet many science communicators on Instagram dissented from Wright's criticisms—including Yammine herself, who co-authored a reply in *Science* with Paige Jarreau and two other colleagues. "Although we agree with [Wright] that there are many systemic structures perpetuating the marginalization of women in science," they wrote, "we view social media as a powerful tool in a larger strategy to dismantle such structures."[35] In particular, they argued that selfies on Instagram help personalize scientists in ways that engage social media users and increase the visibility of women in the profession.

The following year, Jarreau and five of her colleagues—including Yammine—published a study testing how scientists' Instagram selfies influence audience members' perceptions of science.[36] The team had launched the project in 2017, using the hashtag #ScientistsWhoSelfie to help crowdfund their experiment.[37] They found that when scientists posted selfies from their labs or field sites, study participants perceived them as "warmer and more trustworthy, and no less competent."[38] Moreover, the participants who saw selfies of women scientists were less likely to perceive science as being a "male" activity. As Jarreau and her colleagues point out, their findings demonstrate the potential for using Instagram to promote a more positive image of science and to challenge gender stereotypes about the profession.

The results of the Scientists Who Selfie study also highlight how the affordances of specific social media platforms help shape messages on them—and, in turn, the effects of these messages. Instagram's design features encourage the sharing of selfies, but other platforms encourage different types of communication. In the case of Reddit, the affordances of the site foster text-based conversations among laypeople and scientists on its forums, or "subreddits"—some of which revolve around answering users' questions about science.

Asking about Science on Reddit

After high school student Ethan Lindenberger rebelled against his anti-vaxxer mother by getting himself immunized, he publicly lobbied in favor of vaccinations by writing opinion columns, testifying to Congress, and giving television interviews.[39] On April 3, 2019, *Full Frontal* host Samantha Bee asked him where he found his information about vaccines:

LINDENBERGER: So, I turned to Reddit … [Bee makes a horrified face] back in November [2018] and asked about getting vaccinated without my mom's, you know, consent or approval, and everyone's responses were extremely supportive.

BEE: [Incredulously] Sorry, people on Reddit gave you good advice?

LINDENBERGER: Great advice, actually.

BEE: *Reddit?* The *USA Today* of white supremacy—that Reddit?

LINDENBERGER: It just depends on where you go.

Bee and Lindenberger's exchange highlights the multifaceted nature of Reddit. On the one hand, the site has a reputation as a forum for hatemongers, including members of the alt-right.[40] On the other hand, Reddit offers a platform for not only learning about science but also discussing it with both scientists and non-scientists.

Reddit hosts many communities featuring science-themed conversations. One particularly prominent subreddit, r/science, dates back to 2006 and boasted almost 25 million members as of August 2020. Another popular subreddit is r/AskScience, a 20-million member forum where users can ask questions such as "If you melt a magnet, what happens to the magnetism?" and "If we return to the moon, is there a telescope on Earth today strong enough to watch astronauts walking around on the surface?" Reddit also hosts numerous field-based subreddits such as r/biology, r/physics, r/astronomy, and r/chemistry (each of which had one to two million members as of August 2020).

Given that many non-scientists participate in these forums, their moderators often create structures to guide scientific discussions. For example, /r/science enforces an extensive set of rules, the first of which is that all posts "must directly link to recently published peer-reviewed research or media summary."[41] Moreover, the subreddit bans "comments that dispute well-established scientific concepts (e.g., gravity, vaccination, anthropogenic climate change, etc.) [without] appropriate peer-reviewed evidence." Some of the top-voted posts from 2020 shared studies on the function of zebras' stripes, the impact of parenthood on happiness, and the health effects of implementing Medicare for All.

Some Reddit forums also invite scientists to help inform their members about key issues. In 2014, r/science began hosting a series of AMA (Ask Me Anything) discussions where users could interact with scientists such as Stephen Hawking and Neil deGrasse Tyson.[42] When information scientist Noriko Hara and her colleagues interviewed 70 scientists who participated in these AMAs, the respondents reported "overwhelmingly positive" experiences on the site.[43] The moderators of r/science discontinued the series in 2018; since then, however, many scientists have hosted AMAs on the r/AskScience subreddit.

To find out more about people's experiences with science communication on the platform, Jessica McKnight surveyed 135 Reddit users. They reported that commenters on the site often contributed useful new information or corrected misinformation.[44] "If someone makes an incorrect statement, there will be dozens of comments underneath saying why the person is wrong," one respondent said. Another pointed out that the "upvote/downvote system and comments are like peer review." At the same time, the respondents cited several limitations of Reddit as a scientific source, including the frequent use of sensationalized post titles and the difficulty of finding information on particular topics.

As a follow-up, McKnight conducted an experiment testing the effects of Reddit posts on audience members' perceptions of two science-related issues: fracking (hydraulic fracturing, a method of extracting oil or gas from underground rock) and herbal remedies.[45] The fracking post didn't influence audience members' opinions, but the herbal remedies post—which discussed research showing that many of these remedies contain nothing except powdered rice and weeds—did sway participants. In particular, it encouraged those with high science knowledge and interest to view herbal remedies more skeptically.

McKnight's results show that interactions on Reddit help users make sense of science. As such, her findings reinforce Ethan Lindenberger's take on the site. It may be notorious as a font of toxic commentary, but many scientists and laypeople rely on specific subreddits to communicate with one another in ways that can shape users' perceptions of science. Furthermore, these subreddits often implement rules to maintain scientific integrity and legitimacy. Of course, Reddit isn't the only social media community where laypeople generate content or rate and reply to scientific messages. Take YouTube, an even more popular platform with its own distinctive affordances that facilitate video-sharing along with user feedback in form of views, likes, dislikes, and comments.

Sound Waves and Stick Figure Science on YouTube

In 2015, the sixth annual VidCon—billed as "the world's largest online video conference"—featured many of YouTube's most popular hosts as guests.[46] The lineup included not only stars from entertainment and lifestyle channels, but also science communicators such as Michael Stevens of *Vsauce*, Derek Muller of *Veritasium*, Gregory Brown of *AsapSCIENCE*, Vi Hart of *Vihart,* and Emily

Graslie of *The Brain Scoop*.[47] VidCon "is a place where the scientists get just as big screams as the musicians, beauticians, and pranksters," *Scientific American* correspondent Jayde Lovell reported live from the event.[48]

Three years later, our research assistant, D.J. McCauley, took a closer look at the most popular YouTube science channels. She analyzed the 5 most recent videos from the top 30 English-language science channels, each of which had more than 400,000 subscribers (the top channel, *Vsauce*, had 13 million). The most common topics on these channels were physics and astronomy. For example, a video hosted by Dianna Cowern of *Physics Girl* answered the question, "What is a quantum coin toss?" and one from *Minute Physics* used stick-figure drawings to explain the "relativity of simultaneity." Similarly, videos from the animated channel *Kurzgesagt* depicted concepts such as string theory and black holes.

Other channels focused on mathematics, chemistry, biology, or geology. *Periodic Videos* featured entries on cobalt and promethium in the "Periodic Table of Videos" series narrated by Martyn Poliakoff, while Vi Hart of *Vihart* drew her own stick figures to explain mathematical ideas such as metachirality and tau. On *The Brain Scoop*, Emily Graslie talked about the discovery of a fossilized arachnid, and *Naked Science* combined real-life footage with computer animation to illustrate plate tectonics. The most popular video in the sample, however, was from *AsapSCIENCE*. Titled, "Do you hear 'Yanny' or 'Laurel'? (SOLVED WITH SCIENCE)," it drew on principles of acoustics to settle a debate over an auditory illusion.

Collectively, these videos elicited a high level of engagement reflected in information visible to anyone who found them on YouTube. For example, the "'Yanny' or 'Laurel'" video had 33 million views as of May 2018. Many of the other videos had viewing totals in the hundreds of thousands or millions, as well. The "'Yanny' or 'Laurel'" video also had almost half a million upvotes, and a dozen other videos had at least 100,000 upvotes. Downvotes were much less common: the "'Yanny' or 'Laurel'" video had 18 upvotes for every downvote, and only 2 other videos had more than 10,000 downvotes.

These viewing numbers and votes give users signals about how to make sense of science. In particular, they provide what James Spartz and his colleagues call "normative social cues": that is, hints about what audience members should think based on what other people are saying and doing.[49] People often modify their beliefs and behavior to match their peers "in real life," and the same process can play out on social media. To demonstrate this, Spartz and his team manipulated the number of views displayed for a YouTube video about climate change: either more than a million, or less than a thousand. Participants who saw the higher number rated climate change as more important. As the researchers point out, this sort of peer influence could help spark engagement and, ultimately, action on science-related issues.

Nor are viewing numbers, likes, and dislikes the only sorts of social signals available on YouTube. Audience members can also take cues from the number and nature of user comments. The "'Yanny' or 'Laurel'" video had almost 150,000 comments as of May 2018, while 10 other videos in McCauley's sample had at least 10,000 comments. As for the substance of the comments, some engaged with the messages in the videos whereas others tried to hijack those messages to make their own points.

A pair of studies by Matthew Shapiro and Han Woo Park illustrate the latter phenomenon by investigating how YouTube commenters discussed videos about climate change. The first study showed that user comments generally had little to do with the scientific information in the videos themselves. [50] Instead, commenters engaged in "rampant" politicization of climate science through accusations of fraud, alarmism, and manipulation.[51] The second study revealed that a handful of climate activists and skeptics tended to dominate the conversations about each video.[52] Whereas high viewing numbers and likes can lend more weight to scientific messages in YouTube videos, the sorts of narrow and skewed comment threads Shapiro and Park found could discourage users from deliberating thoughtfully about these messages and reinforce polarized perceptions of science.

YouTube commenters sometimes attack the messenger, as well. In particular, Vi Hart and Emily Graslie have described how women science communicators on the site frequently receive sexist comments from users. Hart addresses her experiences with such comments in a January 21, 2013 *Vihart* video titled "Vi Hart's Guide to Comments," and Graslie does the same in a November 27, 2013 *Brain Scoop* video titled "Where My Ladies At?" Which raises the question: can science communicators influence audience members' perceptions by calling out the sexism and stereotyping women often encounter both in scientific workplaces and on social media?

"Where My Ladies At?": Using Social Media to Challenge Sexism in Science

Like offline science, social media science is often a male-dominated space marked by sexism. For example, *Science* magazine's "science stars of Twitter" list included only four women, and the accompanying article quoted one of them—astronomer Pamela Gay—on how sexist responses push women scientists away from social media. "At some point," she explained, "you just get fed up with all the 'why you are ugly' or 'why you are hot' comments."[53] The broader analysis by Qing Ke's team found that though women were better represented on Twitter than they were in scientific publications, the platform still wasn't gender-balanced: only 39% of scientists in their study were women.[54]

YouTube is similarly skewed when it comes to who communicates about science on the site. A 2015 study by Dustin Welbourne and Will Grant found that men outnumbered women as science channel hosts, though the gender of the host didn't affect the viewing numbers for videos.[55] Three years later, D.J. McCauley found that only one in five videos on the top science-themed channels included a woman as a host or co-host.[56]

At the same time, women have used a variety of social media platforms to counter gender stereotypes in science. In 2012, for example, science communicator Allie Wilkinson created the Tumblr site This Is What a Scientist Looks Like to challenge the popular image of a scientist as a "white man in a white lab coat."[57] She asked scientists to submit photographs of themselves and then posted these images along with short biographies.[58] The portraits on Wilkinson's site depicted women and people of color working in a wide range of fields and sites, from a paleoecologist trekking the Austrian Alps to a marine scientist diving off Key Largo, Florida.[59] Tumblr has since declined in popularity, but women scientists and science communicators such as "Science Sam" Yammine now share selfies on Instagram and other social media platforms.

Women have also used hashtags to highlight gender bias in science—as when they called out the sexism reflected in Nobel prize-winning biochemist Tim Hunt's comments at the World Conference for Science Journalists. "Let me tell you about my trouble with girls," he told his audience. "Three things happen when they are in the lab: you fall in love with them, they fall in love with you, and when you criticize them, they cry." Many women scientists responded by tweeting sarcastic commentary along with photographs of themselves and the hashtag #distractinglysexy.[60] "Nothing like a sample tube full of cheetah poop to make you #distractinglysexy," wrote wildlife biologist Sarah Durant. Likewise, marine scientist Michelle LaRue tweeted a photograph of herself in Antarctica and added, "Thank goodness for the cold weather gear, otherwise my male teammates might have fallen in love."[61]

Science communicators have challenged the sexism that plays out on social media, as well. For example, Emily Graslie's "Where My Ladies At?" video discusses the gender-based "internet bullying" she has "had to deal with on a daily basis."[62] The video begins with her observation that more than a dozen science-themed channels hosted by men have at least 400,000 subscribers. "What is preventing women from reaching the same number of people?" she asks. "I feel like women are going to give up more easily because of comments like this." A male co-host then reads some of the sexist comments Graslie has received, such as "'She just needs some sexier glasses,'" "'I'd still totally do her,'" and "'You'd think this was a man's job.'" Sexism and internet harassment in science "are serious issues that need to be discussed," Graslie says at the end of the video. "We need to make … it possible for people of all genders to feel acknowledged for their contributions."

Graslie's "Where My Ladies At?" video reached a wide audience, thanks in part to a publicity boost from coverage by science websites such as ScienceBlogs, feminist websites such as Jezebel, and mainstream news outlets such as National Public Radio. It had received more than a million views as of August 2020. Graslie's channel has also amassed well over 400,000 subscribers, breaking the glass ceiling she highlighted in the video.

In 2015, we conducted an experiment testing the effects of the "Where My Ladies At?" video along with two other *Brain Scoop* videos that didn't explicitly call out sexism.[63] One, titled "Where'd You Get All Those Dead Animals," describes how the Chicago Field Museum of Natural History acquires animal specimens. The other video, "Ask Emily #5," features Graslie answering a series of viewer questions. We randomly assigned 311 undergraduate students to watch one of these videos or (for the control group) no video about science.

The "Where My Ladies At?" video had a clear impact on viewers' perceptions of sexism in science and science communication (Figure 7.2). Compared to participants in the other three groups, those who watched Graslie talk about

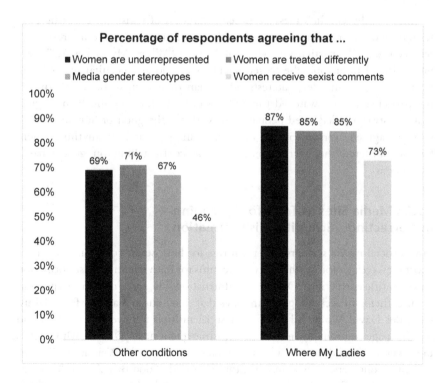

FIGURE 7.2 Perceptions of sexism in science, by *Brain Scoop* watching condition (*The Brain Scoop* Experiment, 2015)

her experiences with gender-based internet bullying were more likely to agree that women are underrepresented in science (87% versus 69%), that women are treated differently than men in science (85% versus 71%), that media portrayals of scientists often include gender stereotypes (85% versus 67%), and that women who communicate publicly about science often receive sexist comments (77% versus 46%).

To be sure, some participants rejected Graslie's message. "Both our culture and probably biology lean women away from STEM at a young age," wrote one man, "so just to be devil's advocate, I'd say ... many women happily choose not to go into STEM, and the statistics of low women in STEM are not entirely due to sexism." However, the "Where My Ladies At?" video resonated with many of the women who took part in the study. "As a woman studying in the STEM field, this really hit home," one wrote. "I can relate to the video," said another. "I have 700k followers on my Vine account and I see sexist comments like this all the time." A third wrote, "PREACH! I'm an average-looking female in the top 5% of my class. I feel I have to speak louder and longer just to be heard or not be reduced to the appeal of my outfit that day."

Importantly, Graslie's message raised awareness of sexism in scientific professions and science communication without creating a negative impression of science itself. The participants who watched the "Where My Ladies At?" were no likelier than anyone else to see scientists as odd and peculiar, or as being no fun, or as having few interests other than their own work. Furthermore, the participants who watched this video stood out for how much they *agreed* that scientists are dedicated people who work for the good of humanity (37% strongly agreed, versus 15% in the other three groups). If anything, then, Graslie bolstered the overall image of science by taking on gender-based harassment.

Social Media Sites as Tools for Spreading— or Correcting—Scientific Misinformation

Just as social media sites provide platforms for both sexist comments and anti-sexist messages, so too do they serve as forums for both scientific misinformation and scientific corrections. On the misinformation side, some of the most popular science-themed Facebook pages are ones that share questionable or false claims. Consider David Wolfe, who had almost 12 million followers as of 2020 and uses his page to promote anti-vaccine misinformation. Or Dr. Oz, who pitches unproven health fads such as a "7-Day Grapefruit Detox for Weight Loss" to his 6 million followers.[64] Or Vani Hari, the self-titled "Food Babe," who has more than a million Facebook followers and shares misleading information about genetically modified foods.[65]

Nor is Facebook the only social media platform for scientific misinformation. A CNN investigation during the midst of a 2019 measles outbreak discovered that Instagram searches for "vaccines" yielded numerous posts with false claims and hashtags such as #VaccinesKill.[66] Similarly, a 2020 study by the nonprofit Avaaz found that 16% of the top 100 YouTube search results for "global warming" led to videos featuring misinformation about climate change.[67] And when the COVID-19 virus spread across the world, conspiracy theories about the origins of the pandemic and misinformation about unproven "cures" such as hydroxychloroquine circulated on every major social media site.[68]

A rumor about Bill Gates illustrates how false information can travel from social media platforms to news outlets and then back to social media. The seed of the rumor came from a March 18, 2020 "Ask Me Anything" that Gates hosted on Reddit, where he fielded questions about COVID-19 and public health responses to the pandemic.[69] In answering one question, he wrote, "Eventually we will have some digital certificates to show who has recovered or been tested recently or when we have a vaccine who has received it." The next day, a relatively obscure website, Biohacking, posted a story titled, "Bill Gates will use microchip implants to fight coronavirus."[70] Gates was actually referring to electronic health documents, not microchips, but Biohacking's YouTube video about its conspiracy theory went viral and received millions of views.[71] The following month, former Donald Trump advisor Roger Stone mentioned the rumor in a radio interview. When the *New York Post* covered his comments, Facebook users shared the newspaper's story more than a million times.

Public outcries over scientific misinformation on social media have prompted tech companies to work on strategies for correcting such misinformation and limiting its spread. In 2019, Google announced that it would demonetize anti-vaccine videos on YouTube (which it owns) by removing ads from them.[72] That same year, Facebook pledged to block anti-vaccine hashtags on Instagram (which it owns) and added information pop-ups on Facebook that linked users searching for "vaccines" to the website of the Centers for Disease Control (CDC).[73] Meanwhile, Twitter directed users searching for tweets about vaccines to a post from the US Department of Health and Human Services.[74]

During the COVID-19 pandemic, both Google and Facebook tried to block false claims about the virus and share scientifically sound public information with social media users.[75] In May 2020, Twitter went a step further by adding fact-checks to misleading tweets.[76] Yet these companies struggled to contain the flow of misinformation on their social media platforms. For example, the "Plandemic" conspiracy theory video resurfaced on YouTube and Facebook even after the sites banned it.[77]

Studies by Leticia Bode and Emily Vraga show that efforts to correct scientific misinformation on social media can work—at least some of the time. In one experiment, they found that Facebook links to "related articles" debunking false claims about the health effects of GMOs reduced participants' misperceptions on the topic.[78] By contrast, stories debunking the notion that the MMR vaccine causes autism had no effect on audience members. As Bode and Vraga argue, this false claim has circulated so widely that people who already believe it may argue away any messages to the contrary.

On a more optimistic note, Bode and Vraga's follow-up studies have shown that other approaches to correcting scientific misinformation on social media can also work. For example, a second experiment found that sharing information from an expert source—in this case, the CDC—was particularly effective in debunking a conspiracy theory that blamed the Zika virus on genetically modified mosquitos.[79] Similarly, a third study showed that two different forms of correction helped dispel belief in the same conspiracy theory: links generated by Facebook algorithms and stories shared by Facebook users.[80]

A national survey Bode and Vraga conducted during the COVID-19 pandemic reinforces the point that social media users can help fact-check false information. Around one-third of the respondents reported seeing someone else being told they'd shared misinformation about the virus, and almost a quarter said they'd corrected such misinformation themselves.[81] Moreover, large majorities agreed that it's important for social media users to help correct misinformation and that it's everyone's responsibility to do so.

Judging by these results, the best approach to challenging false scientific claims on social media may be a multi-pronged one. Tech companies can use algorithms to show corrections; experts can use lend their credibility to debunk faulty science, and ordinary users can speak up when they see misinformation.

Social Media Science: Pitfalls and Promising Directions

Much like film, television, and print outlets, social media platforms present a wide range of messages about science. At one end of the spectrum, scientists use these sites to share research with one another and the broader public. At the other end of the spectrum, figures such as David Wolfe and Vani Hari rely on the same platforms to spread bogus claims. Meanwhile, science communicators such as Elise Andrew, Samantha "Science Sam" Yammine, and Emily Graslie provide a blend of information and entertainment that can spark enthusiasm for science, enhance the image of scientists, and challenge sexism in scientific professions.

Unlike other forms of media, social media sites make it easy for users to create their own content and reply to messages immediately and publicly. Sometimes these interactions carry baneful consequences—as when users bully science communicators with sexist comments. In other instances, however, users provide valuable scientific information—as when "Redditors" pointed Ethan Lindenberger to sound advice on vaccinations. Users can also play important roles in correcting misinformation on social media.

More broadly, social media platforms offer new opportunities for scientists and non-scientists to interact with one another in ways that enhance public faith in and engagement with science. Highlighting this potential, one recent cross-national study by Brigitte Huber and her colleagues found that social media users were especially likely to trust science.[82] Still, these sorts of effects may depend on the algorithms of social media platforms, the social networks of users, and the effectiveness of science communicators at engaging with audiences.[83] No one messaging approach will work for all sites or all users, but Facebook pages such as IFLS, Twitter accounts of scientists such as Pamela Gay, Instagram accounts such as @sciencesam, Reddit forums such as r/AskScience, and YouTube channels such as *The Brain Scoop* illustrate the promise of social media science.

Notes

1 Fitts, Alexis Sobel, "Do you know Elise Andrew?" *Columbia Journalism Review*, September/October 2014, https://archives.cjr.org/cover_story/elise_andrew.php; Teeman, Tim, "Why millions love Elise Andrew's science page," *Guardian*, October 13, www.theguardian.com/science/2013/oct/13/i-fucking-love-science-elsie-andrew.

2 Sturgess, Kylie, "How much do you love science? Interview with Elise Andrew," *Skeptical Inquirer*, July 1, 2013, https://skepticalinquirer.org/exclusive/how-much-do-you-love-science-interview-with-elise-andrew.

3 Fitts, "Do you know Elise Andrew?"

4 Pomeroy, Ross, "The worst kind of science clickbait," *RealClearScience*, May 8, 2017, www.realclearscience.com/blog/2017/05/08/the_worst_kind_of_science_clickbait.html.

5 Hitlin, Paul, and Kenneth Olmstead, "The science people see on social media," Pew Research Center, March 21, 2018, www.pewresearch.org/science/2018/03/21/the-science-people-see-on-social-media.

6 Funk, Cary, Jeffrey Gottfried, and Amy Mitchell, "Science news and information today," Pew Research Center, September 20, 2017, www.journalism.org/2017/09/20/science-news-and-information-today.

7 See the Appendix for details.

8 Brossard, Dominique, and Dietram A. Scheufele, "Science, new media, and the public," *Science* 339, no. 6115 (2013): 40–41; Lee, Nicole M., and Matthew S. VanDyke, "Set it and forget it: The one-way use of social media by government agencies communicating science," *Science Communication* 37, no. 4 (2015): 533–541.

9 Tufekci, Zeynep, "Big questions for social media big data: Representativeness, validity and other methodological pitfalls," *arXiv preprint arXiv:1403.7400* (2014), 2–3; Ley, Barbara L., and Paul R. Brewer, "Social media, networked protest, and the March for Science," *Social Media+ Society* 4, no. 3 (2018): 2056305118793407.

10 Spartz, James T., Leona Yi-Fan Su, Robert Griffin, Dominique Brossard, and Sharon Dunwoody, "YouTube, social norms and perceived salience of climate change in the American mind," *Environmental Communication* 11, no. 1 (2017): 1–16.

11 Hitlin and Olmstead, "The science people see."

12 See the Appendix for details.

13 IFLScience, February 21, 2020, www.facebook.com/IFLScience/posts/3422615091092811.

14 Hall, Neil, "The Kardashian index: A measure of discrepant social media profile for scientists," *Genome Biology* 15, no. 7 (2014): https://doi.org/10.1186/s13059-014-0424-0.

15 Liang, Xuan, Leona Yi-Fan Su, Sara K. Yeo, Dietram A. Scheufele, Dominique Brossard, Michael Xenos, Paul Nealey, and Elizabeth A. Corley, "Building buzz: (Scientists) communicating science in new media environments," *Journalism & Mass Communication Quarterly* 91, no. 4 (2014): 772–791.

16 You, Jia, "Who are the science stars of Twitter?" *Science* 345, no. 6203 (2014): 1440–1441.

17 Neil deGrasse Tyson, March 13, 2020, https://twitter.com/neiltyson/status/1238564845781385217?s=20.

18 Neil deGrasse Tyson, February 26, 2020, https://twitter.com/neiltyson/status/1232841860386709506?lang=en.

19 Coleman, Nancy, "Neil deGrasse Tyson's tweet on mass shooting deaths strikes a nerve," *New York Times*, August 5, 2019, www.nytimes.com/2019/08/05/arts/neil-degrasse-tyson-el-paso.html.

20 Cox, Brian, July 1, 2018, https://twitter.com/profbriancox/status/1013545554121576448?lang=en.

21 Elmhirst, Sophie, "Is Richard Dawkins destroying his reputation?" *Guardian*, June 9, 2015, www.theguardian.com/science/2015/jun/09/is-richard-dawkins-destroying-his-reputation.

22 Anita Singh, "Richard Dawkins in storm over 'mild date rape' tweets," *Telegraph*, July 29, 2014, www.telegraph.co.uk/news/uknews/10998498/Richard-Dawkins-in-storm-over-mild-date-rape-tweets.html.

23 Richard Dawkins, February 16, 2020, https://twitter.com/RichardDawkins/status/1228943686953664512?s=20.

24 Ke, Qing, Yong-Yeol Ahn, and Cassidy R. Sugimoto, "A systematic identification and analysis of scientists on Twitter," *PloS one* 12, no. 4 (2017): e0175368.

25 McQuate, Sarah, "What all those scientists on Twitter are really doing," *Nature News*, April 20, 2017, www.nature.com/news/what-all-those-scientists-on-twitter-are-really-doing-1.21873.

26 Jarreau, Paige Brown, Imogene A. Cancellare, Becky J. Carmichael, Lance Porter, Daniel Toker, and Samantha Z. Yammine, "Using selfies to challenge public stereotypes of scientists," *PloS one* 14, no. 5 (2019): e0216625.

27 Trackalytics, "The most followed Instagram profiles," August 29, 2020, www.trackalytics.com/the-most-followed-instagram-profiles/page/1.

28 *NewsWhip*, "Interview: why NASA's social media is out of this world," *NewsWhip*, March 27, 2017, www.newswhip.com/2017/03/nasa-social-media-interview.

29 Trackalytics, "Most followed Instagram profiles."

30 CBS News, "National Geographic's most popular Instagram photos," CBS News, September 4, 2020, www.cbsnews.com/pictures/national-geographics-most-popular-instagram-images.

31 Amaria, Kainaz, "National Geographic hit 100 million Instagram followers. To celebrate, it wants your images for free," *Vox*, February 2, 2019, www.vox.com/2019/2/20/18232426/national-geographic-100m-instagram-followers-image-contest-free.

32 Jarreau et al., "Using selfies."

33 Jarreau et al., "Using selfies"; Pew Research Center, "Social media fact sheet," Pew Research Center, June 12, 2019, www.pewresearch.org/internet/fact-sheet/social-media/#who-uses-each-social-media-platform.

34 Wright, Meghan, "Why I don't use Instagram for science outreach," *Science*, February 2, 2019, www.sciencemag.org/careers/2018/03/why-i-dont-use-instagram-science-outreach.

35 Yammine, Samantha Z., Christine Liu, Paige B. Jarreau, Imogen R. Coe, "Social media for social change in science," *Science*, April 13, 2018, https://science.sciencemag.org/content/360/6385/162.2?source=post_page.

36 Jarreau et al., "Using selfies."

37 Qaiser, Farah, "Scientists are fostering public trust on social media, one selfie at a time," *Massive Science*, May 6, 2019, https://massivesci.com/articles/scientists-selfie-instagram-public-trust-social-media.

38 Jarreau et al., "Using selfies."

39 Lindenberger, Ethan, "Growing up unvaccinated: My anti-vaxx mother made me a health risk for the whole community," *USA Today*, February 26, 2019, www.usatoday.com/story/opinion/voices/2019/02/26/measles-mumps-vaccines-teens-autism-outbreak-column/2978717002.

40 Marantz, Andrew, "Reddit and the struggle to detoxify the internet," *New Yorker*, March 12, 2018, www.newyorker.com/magazine/2018/03/19/reddit-and-the-struggle-to-detoxify-the-internet; Romano, Aja, "Reddit just banned one of its most toxic forums. But it won't touch The_Donald," *Vox*, November 11, 2017, www.vox.com/culture/2017/11/13/16624688/reddit-bans-incels-the-donald-controversy; Statt, Nick, "Reddit CEO says racism is permitted on the platform, and users are up in arms," *The Verge*, April 11, 2018, www.theverge.com/2018/4/11/17226416/reddit-ceo-steve-huffman-racism-racist-slurs-are-okay.

41 Reddit, "r/science rules," September 4, 2020, www.reddit.com/r/science.

42 Reddit, "r/science will no longer be hosting AMAs," September 4, 2020, www.reddit.com/r/IAmA/comments/8khv45/rscience_will_no_longer_be_hosting_amas.

43 Hara, Noriko, Jessica Abbazio, and Kathryn Perkins, "An emerging form of public engagement with science: Ask Me Anything (AMA) sessions on Reddit r/science," *PloS One* 14, no. 5 (2019): e0216789, 14.

44 McKnight, Jessica, "'The new Reddit journal of science': Public evaluation and understanding of scientific information based on source factors in social media," MA thesis, University of Delaware, 2015, 14.

45 McKnight, "'The new Reddit journal,'" 14.

46 Lovell, Jayde, "YouTube's rock stars of science make a splash a VidCon," *Scientific American*, July 30, 2015, https://blogs.scientificamerican.com/guest-blog/youtube-s-rock-stars-of-science-make-a-splash-a-vidcon1.

47 DeSimone, "VidCon 2015 featured creators: Check out this year's VidCon Lineup," *New Media Rock Stars*, April 21, 2015, https://newmediarockstars.com/2015/04/vidcon-2015-featured-creators-check-out-this-years-vidcon-lineup.

48 Lovell, "YouTube's rock stars."

49 Spartz et al., "YouTube, social norms, and perceived salience."

50 Shapiro, Matthew A., and Han Woo Park, "More than entertainment: YouTube and public responses to the science of global warming and climate change," *Social Science Information* 54, no. 1 (2015): 115–145.

51 Shapiro and Park, "More than entertainment," 13.

52 Shapiro, Matthew A., and Han Woo Park, "Climate change and YouTube: Deliberation potential in post-video discussions," *Environmental Communication* 12, no. 1 (2018): 115–131.

53 You, "Who are the science stars."

54 Ke et al., "A systematic identification."

55 Welbourne, Dustin J., and Will J. Grant, "Science communication on YouTube: Factors that affect channel and video popularity," *Public Understanding of Science* 25, no. 6 (2016): 706–718.

56 Nor was YouTube science notably diverse in terms of race. The communicators on the top channels were almost exclusively white, and *Science Plus* was the lone channel in the sample to feature a woman of color as a host.

57 DiChristina, Marietta, "What a scientist looks like," *Scientific American*, February 16, 2012, www.scientificamerican.com/article/what-a-scientist-looks-like.

58 Wilcox, Christie, "This is what a scientist looks like," *Discover*, February 16, 2012, www.discovermagazine.com/the-sciences/this-is-what-a-scientist-looks-like.

59 Wilkinson, Allie, "This is what a scientist looks like," September 4, 2020, https://lookslikescience.tumblr.com.

60 Bilefsky, Dan, "Women respond to Nobel laureate's 'trouble with girls,'" *New York Times*, June 11, 2015, www.nytimes.com/2015/06/12/world/europe/tim-hunt-nobel-laureate-resigns-sexist-women-female-scientists.html.

61 Chappell, Bill, "#Distractinglysexy tweets are female scientists' retort to 'disappointing' comments," National Public Radio, June 12, 2015, www.npr.org/sections/thetwo-way/2015/06/12/413986529/-distractinglysexy-tweets-are-female-scientists-retort-to-disappointing-comments.

62 Graslie, Emily. "Where my ladies at?" *The Brain Scoop with Emily Graslie*, November 27, 2013, www.youtube.com/watch?v=yRNt7ZLY0Kc.

63 For more details about this study, see Brewer, Paul R., and Barbara L. Ley, "'Where my ladies at?': Online videos, gender, and science attitudes among university students," *International Journal of Gender, Science and Technology* 9, no. 3 (2018): 278–297.

64 Hall, Harriet A., "The incorrigible Dr. Oz," SkepDoc, August 29, 2017, www.skepdoc.info/the-incorrigible-dr-oz.

65 Godoy, Maria, "Is the Food Babe a fearmonger? Scientists are speaking out," National Public Radio, December 4, 2014, www.npr.org/sections/thesalt/2014/12/04/364745790/food-babe-or-fear-babe-as-activist-s-profile-grows-so-do-her-critics.

66 Yurieff, Kaya, "Instagram still doesn't have vaccine misinformation under control," CNN Business, May 8, 2019, www.cnn.com/2019/05/08/tech/instagram-vaccine-misinformation/index.html.

67 Avaaz, "Why is YouTube broadcasting climate misinformation to millions?" Aavaz, January 16, 2020, https://secure.avaaz.org/campaign/en/youtube_climate_misinformation.

68 Kafka, Peter, "Big Tech thought the pandemic wouldn't be political. Think again," *Vox*, May 27, 2020, www.vox.com/recode/2020/5/27/21270280/facebook-twitter-youtube-coronavirus-pandemic-misinformation-political-controversy-face-masks.

69 Gates, Bill, "I'm Bill Gates, co-chair of the Bill & Melinda Gates Foundation. AMA about COVID-19," Reddit, March 18, 2020, www.reddit.com/r/Coronavirus/comments/fksnbf/im_bill_gates_cochair_of_the_bill_melinda_gates.

70 Lytvynenko, Jane, "Here's a timeline of how a Bill Gates Reddit AMA turned into a coronavirus vaccine conspiracy," *BuzzFeed*, April 18, 2020, www.buzzfeednews.com/article/janelytvynenko/conspiracy-theorists-are-using-a-bill-gates-reddit-ama-to.

71 Ball, Philip, and Amy Maxmen, "The epic battle against coronavirus misinformation and conspiracy theories," *Nature*, May 27, 2020, www.nature.com/articles/d41586-020-01452-z; Reuters Fact Check, "False claim: Bill Gates planning to use microchip implants to fight coronavirus," March 31, 2020, www.reuters.com/article/uk-factcheck-coronavirus-bill-gates-micr/false-claim-bill-gates-planning-to-use-microchip-implants-to-fight-coronavirus-idUSKBN21I3EC.

72 Shu, Catherine, "YouTube demonetizes anti-vaccination videos," *TechCrunch*, February 22, 2019, https://techcrunch.com/2019/02/22/youtube-demonetizes-anti-vaccination-videos.

73 Belluz, Julia, "Facebook, Pinterest, and YouTube are cracking down on fake vaccine news," *Vox*, September 5, 2019, www.vox.com/2019/3/1/18244384/measles-outbreak-vaccine-washington; Birnbaum, Emily, "Instagram to block anti-vaccine hashtags amid misinformation crackdown," *The Hill*, March 21, 2019, https://thehill.com/policy/technology/435207-instagram-to-block-anti-vaccine-hashtags-amid-misinformation-crackdown.

74 Kelly, Makena, "Twitter fights vaccine misinformation with new search tool," *The Verge*, May 14, 2019, www.theverge.com/2019/5/14/18623494/twitter-vaccine-misinformation-anti-vax-search-tool-instagram-facebook.

75 O'Sullivan, Donie, "How Covid-19 misinformation is still going viral," CNN Business, May 9, 2020, www.cnn.com/2020/05/08/tech/covid-viral-misinformation/index.html.

76 Dayaram, Sareena, "Twitter fact-checks China official's post claiming coronavirus originated in US," *CNET*, May 28, 2020, www.cnet.com/news/twitter-fact-checks-china-officials-post-claiming-coronavirus-originated-in-us.

77 O'Sullivan, "How Covid-19 misinformation is still going viral."

78 Bode, Leticia, and Emily K. Vraga, "In related news, that was wrong: The correction of misinformation through related stories functionality in social media," *Journal of Communication* 65, no. 4 (2015): 619–638.

79 Vraga, Emily K., and Leticia Bode, "Using expert sources to correct health misinformation in social media," *Science Communication* 39, no. 5 (2017): 621–645.

80 Bode, Leticia, and Emily K. Vraga, "See something, say something: Correction of global health misinformation on social media," *Health Communication* 33, no. 9 (2018): 1131–1140.

81 Bode, Leticia, and Emily Vraga, "Americans are fighting coronavirus misinformation on social media," *Washington Post*, May 7, 2020, www.washingtonpost.com/politics/2020/05/07/americans-are-fighting-coronavirus-misinformation-social-media.

82 Huber, Brigitte, Matthew Barnidge, Homero Gil de Zuniga, and James Liu, "Fostering public trust in science: The role of social media," *Public Understanding of Science* 28, no. 7 (2019): 759–777.

83 Howell, Emily, and Dominique Brossard, "Science engagement and social media: Communicating across interests, goals, and platforms," in *Theory and Best Practices in Science Communication Training*, ed. Todd P. Newman, Routledge, 2019, 57–70.

8

FORENSIC SCIENCE

Gil Grissom: I tend not to believe people. People lie. The evidence doesn't lie.
— *CSI: Crime Scene Investigation* (2000, Season 1, Episode 3: Crate 'n' Burial)

John Oliver: On TV and in real life, forensic science plays an important role in criminal convictions. Prosecutors often complain about a so-called "*CSI* effect" where jurors expect to see forensic evidence in every case. The problem is, not all forensic science is as reliable as we've become accustomed to believe. A report in 2009 by the National Academy of Sciences found that many forensic sciences do not meet the fundamental requirements of science.
— *Last Week Tonight with John Oliver* (October 1, 2017)

In the summer of 2005, the two of us visited Las Vegas for the first time. We didn't have a Vegas wedding (we'd gotten married two years earlier), or gamble (much, though one of us had some luck with the slots), or see any shows (no David Copperfield, Blue Man Group, or Cirque du Soleil for us). Instead, we wandered past the casinos during the day: the black glass pyramid, the giant Fisher-Price castle, the fake Italian villa with the fountains, and so on. One evening, we were so tired that we stayed in our Hotel Tropicana room and watched our first few episodes of the hit CBS television series *CSI: Crime Scene Investigation*.

By that point, the program had already completed its fourth season and was one of the top-rated prime-time television programs in the United States.[1] In the episodes we watched, crime lab supervisor Gil Grissom (trained as an

DOI: 10.4324/9781003190721-8

entomologist) and his team of investigators—including Catherine Willows, Warrick Brown, Nick Stokes, and Sara Sidle—used DNA, fingerprints, traces of hair or fiber, and other forms of forensic evidence to find the culprits in various unusual crimes (usually gruesome murders) committed in Las Vegas.

Because we are, in some ways, stereotypical science nerds—social science nerds, that is—we immediately began discussing how watching *CSI* might influence audience members. At first, we joked about how the spectacularly high murder rate on the show might lead viewers to see Las Vegas as the most dangerous place in the United States. After all, George Gerbner and other cultivation theorists have argued that television portrayals of violent crime can fuel a "mean world syndrome" in which viewers come to see the real world as resembling the scary one on prime time.[2] From there, however, we turned to talking about another possibility: that *CSI* and programs like it might influence viewers' perceptions of forensic science and scientists. For example, would watching Grissom and his team use DNA evidence to solve cases boost public faith in the reliability of DNA testing? Would the depictions of forensic techniques on *CSI* shape—and potentially distort—perceptions of how widely used, and how swift, such approaches are in real-world criminal investigations? And would the show's portrayals of heroic crime scene investigators foster a positive image of forensic science as a profession?

After our trip, we discovered that other people were already asking the same questions. As early as 2002, *TIME* magazine writer Jeffrey Kluger had raised the prospect of a "*CSI* effect" whereby exposure to the program's portrayals of forensic science might "poison" jurors' responses to forensic evidence.[3] Over the next few years, media outlets ranging from *Scientific American* and *National Geographic* to *USA Today* speculated about the program's impact, including its potential effects not only in the courtroom but also among students searching for careers and even criminals seeking to avoid detection by "cleaning" the scenes of their crimes.[4]

Against this background, we started investigating how *CSI* portrays forensic science, along with whether—and if so, how—media messages shape public perceptions of forensic evidence. As fate would have it, one of us was called for jury duty while we were conducting our research. During the jury selection process, the prosecutor warned potential jurors that the trial at hand would *not* revolve around DNA evidence—unlike many cases on *CSI*. We suspect he said this to guard against the possibility that jurors who watched the show would hold misplaced expectations about the prosecution presenting such evidence.

Reflecting such concerns, along with mirror-image concerns among defense attorneys, part of our research here has focused on the potential for television shows about criminal investigations to influence how viewers weigh DNA evidence and other forms of forensic science in courtroom settings. Yet we've also looked beyond the *CSI* effect (or effects) to investigate whether other forms of media use—including overall television viewing and news media consumption—help explain people's views about forensic science. In addition to swaying perceptions about the scientific credibility of forensic techniques, media messages could influence views on public policies related to criminal investigations. For example, these messages could shape support for resources such as crime labs and DNA databanks. Likewise, messages in both entertainment media and news media might influence audience members' broader understandings of DNA—which, in turn, could spill over to perceptions of medical genetic testing, ancestry testing, and other uses of DNA analysis.

Understanding the role of media messages in shaping perceptions of these topics is important for its own sake, given the increasing prevalence of forensic methods in the criminal justice system and society more broadly. At the same time, investigating the case of the *CSI* effect also gives us an opportunity to explore the bigger question of how media messages help mold public beliefs about *scientific authority*—that is, beliefs about who gets to speak for science and what counts as scientific. Entertainment and news media portrayals of many different fields, from biotechnology and climatology to "fringe" disciplines such as parapsychology, use images, language, and storylines that could bolster or undermine audience perceptions of how credible these fields are. Media depictions may even shape views about the reliability of specific techniques within a given field. In forensic science, for instance, scientific and legal experts attach greater weight to some forms of evidence (such as DNA testing) than other forms of evidence (such as analysis of trace evidence). So, do audience members make the same distinctions?

Our first step in exploring these topics is to track the rise of entertainment media messages about forensic science, from the 19th-century creation of the detective genre to the post-2000 proliferation of forensic-themed television shows, including *CSI* itself, its spinoffs (*CSI: Miami*, *CSI: New York*, and the short-lived *CSI: Cyber*), its "clones" (such as *Without a Trace*, *Cold Case*, *NCIS*, and *Bones*), and its reality-based counterparts (such as *Forensic Files*).

Forensic Science on Prime-Time TV

Producer Jerry Bruckheimer may have transformed the prime-time landscape when he developed *CSI*, but he didn't invent the genre of stories about investigators who use forensic evidence to solve crimes. As far back as 1841, Edgar Allan Poe wrote "tales of ratiocination" such as "The Murders in the Rue Morgue," in which detective C. Auguste Dupin uses a tuft of orangutan hair to help solve a grisly crime. Half a century later, Arthur Conan Doyle created the most famous fictional detective of all: Sherlock Holmes, who is a master at inferring details about crimes from evidence such as soil samples, tobacco ash, gunshot residue, and prints (finger, foot, or even the paws of a gigantic hound). The popularity of the genre continued through the rise of the film industry with classics such as 1935's *The 39 Steps* and later carried over to television in the form of programs such as *Quincy, M.E.*, a drama about a Los Angeles medical examiner that aired from 1976 to 1983.[5]

When *CSI* premiered in 2000, it blended these genre elements with computer-generated special effects and graphic imagery of corpses. The show also emulated the hit medical drama *E.R.* by following the advice of expert consultants—in this case, forensic scientists—on how to integrate high-tech equipment, professional jargon, and real-world procedures into its plots.[6] The combination evidently appealed to viewers. By its second season, *CSI* was one of the ten most-watched programs in the United States, a status it would maintain for the next seven seasons. It also became a broader cultural phenomenon, spawning tie-in books, video games, and a traveling museum exhibit titled "The *CSI* Experience."

Given the show's popularity and the growing debate over its potential effects, researchers from a variety of fields began analyzing its messages about forensic science. One study, by Kimberlianne Podlas, showed that an overwhelming majority of the episodes from the show's first two seasons featured at least one form of forensic evidence.[7] Another study, by Steven Smith and his colleagues, found that the first 2 seasons of *CSI* and the spinoff *CSI: Miami* featured more than 75 types of forensic evidence, with DNA being the most common one—and that the program's heroes solved 97% of their cases.[8]

In portraying forensic evidence, *CSI* uses visuals and dialogue that convey an aura of realism. For example, Sarah Keturah Deutsch and Gray Cavender found that episodes from the first season of the program used sets (sleek-looking labs), props (beakers full of colorful liquids along with microscopes and spectrometers), and jargon ("ulnar loops" in fingerprints analysis, "luminol" in searching for blood) to make the show's science seem plausible.[9] These depictions don't necessarily match how forensic science looks or sounds in the real world; what's more important is that they match audience members' expectations and create a sense of *perceptual realism*.[10] By doing so, *CSI* helps viewers suspend disbelief about the use of forensic science on the program—and, perhaps, in real life as well.[11]

Building on this research, we took a closer look at how *CSI* portrays DNA testing, the most frequently used forensic technique on the program and the most trusted one in real life. Forensic scientists and legal experts regard DNA evidence as the "gold standard" in terms of its reliability.[12] More broadly, the growing focus on DNA in forensic science mirrors the late 20th-century emergence of genetics as a dominant framework for understanding human health, identity, and behavior.[13] In keeping with this, the use of DNA in criminal investigations has expanded dramatically over the past few decades. The Combined DNA Index System (CODIS) unit of the US Federal Bureau of Investigation (FBI) has used the National DNA Index to assist in more than half a million investigations since its founding in 1990.[14] Starting in 1989, organizations such as the Innocence Project have also used forensic DNA testing to exonerate hundreds of wrongly convicted persons, including some awaiting executions on death row.[15]

To capture messages about DNA on *CSI*, we conducted a content analysis of 51 episodes randomly selected from the program's first 7 seasons, a period covering the peak of its popularity.[16] Taken together, these episodes include 82 separate cases. For each case, we coded every step in the process of finding and analyzing DNA. We also examined what characters say about DNA and how the show portrays its forensic scientist protagonists.

DNA evidence plays a prominent role in the episodes we studied. The show's characters search for DNA from unknown sources—usually in the form of hairs, skin, or bodily fluids at the crime scene—in two-thirds of all cases (66%), and they almost always find it when they search for it (93% of the time). Furthermore, they take DNA samples from a known source—typically a suspect or relative—in around half of all cases (52%). Once the investigators collect a DNA sample, they almost always sequence it (84% of the time for DNA from unknown sources; 98% of the time for DNA from known sources)—and in the vast majority of these cases, they find a match (88%). Such results prove highly useful, with more than half of all matches (57%) contributing to the solving of the case.

All told, the CSIs solve 88% of their cases, and DNA matches help them solve 29% of the cases they crack. Given that some episodes include more than one case, the percentage of *episodes* in which DNA evidence helps solve a case is even higher, at 39%. Sometimes matches merely help the heroes rule out a suspect. In other cases, however, a match serves to identify the culprit: "It's her DNA, case closed," to quote a lab technician in one episode.[17] There are only two cases where investigators make mistakes in interpreting DNA evidence. One involves a suspect who has an identical twin brother—a fact the CSIs eventually figure out.[18] The other involves the main characters using newer, more sophisticated techniques to correct a past wrongful conviction based on DNA evidence.[19]

The episodes we analyzed not only portray DNA testing as a common, useful, and highly reliable tool for solving crimes; they also present such testing as strikingly swift. In one episode, Grissom's team collects and analyzes multiple DNA samples in 24 hours to meet a deadline for a trial.[20] In other episodes, the investigators deliver DNA samples and then receive the results of DNA tests a few scenes later, before they even change clothes (so, presumably that same day). Furthermore, the program's sole DNA lab technician seems to handle all the testing in-house without much difficulty. Even when the program touches on backlogs in DNA testing—a common real-world issue—the investigators simply order the technician to "suck it up" or to make "this ... the only case you work on," which evidently solves the problem.[21]

Aesthetically, *CSI* presents DNA analysis as both *credible* and *cool*. The aura of credibility comes from the show's use of dialogue ("epithelial," "epidermal tissue," "markers") and visuals (the lab, with its pipettes, test tubes, and clinical centrifuge) to create perceptual realism. The coolness, in turn, comes from the blue lighting and pulsing music that typically accompanies lab scenes, along with the playful and sometimes sexy banter between the investigators. For example, when a lab technician asks Catherine Willows whether she ever wore a headdress during her days as an exotic dancer, she tells him, "I wore nothing but skin."[22]

Beyond the lab, *CSI* dramatizes the role of forensic science in the judicial process and of DNA in society. A recurring theme on the show is how public officials, prosecutors, and jurors expect forensic evidence. In one episode, an investigator observes that "it is better to have one piece of forensic evidence than ten witnesses"; in another, a character says, "For a conviction, we're going to need a DNA sample."[23] The characters also discuss broader topics involving DNA, including its discovery by James Watson and Francis Crick, the use of DNA testing to establish biological parentage, and the extent to which genes determine one's identity.

In addition to depicting forensic science itself, *CSI* features character portrayals that could sway viewers' perceptions of what scientists are like and who can become one. For the most part, the show's scientists are defined by their intelligence, courage, likability, and dedication to justice (along with their Hollywood looks). Rather than being eccentric loners, they work as a team, supporting— and sometimes rescuing—one another. As with most contemporary prime-time heroes, they do have personal and professional flaws. For example, the pilot episode highlights Warrick Brown's gambling addiction, while another episode shows a technician using the crime lab to analyze his date's DNA ("What I need to know is on the inside, and let me tell you—this girl has got some fine epithelials").[24] In terms of who the *CSIs* are, team leader Gil Grissom checks many of the boxes for the standard stereotype of a scientist: he's white, male, middle-aged, and quirky, with a special fondness for bugs (not to mention a penchant for groan-inducing quips). Yet his colleagues are diverse in terms of sex and race: two of the show's original leads are women, and one is a Black man. They diverge from stereotypes of scientists in other ways, too; as a case in point, Catherine Willows is a single mother.

Looking at the portrayals on *CSI*, it seems reasonable to suspect that watching the program might influence audience members in several different ways. For example, its frequent discussion of DNA could lead viewers to believe they understand the science of DNA testing. Moreover, the show's depiction of DNA testing as virtually infallible could lead them to trust in its reliability and support the use of DNA databanks. At the same time, *CSI*'s portrayals of DNA analysis as routine and swift might fuel unrealistic expectations about its uses in real-world criminal investigations—just the sort of effect prosecutors have worried about. More broadly, the show's messages about DNA could shape viewers' understandings of genetics and its social implications by reinforcing the prominence and "mystique" of genomic science in contemporary culture.[25] Building on social cognitive theory, which suggests that media models can influence viewers' own aspirations and behaviors, it also seems plausible that *CSI*'s portrayal of an appealing and diverse scientific team would help foster interest in forensic science programs.[26]

At the very least, the popularity of Gil Grissom and his team had one clear effect: it sparked the rise of more television shows about forensic science.

The Spinoffs and Clones of *CSI*

Once *CSI* rose toward the top of the Nielsen ratings, CBS greenlit two spinoffs: *CSI: Miami*, which premiered in 2002, and *CSI: New York*, which debuted in 2004. The former aired for ten seasons and cracked the Nielsen top ten in three of those seasons, while the latter aired for nine seasons. Both programs replicated the original's focus on the use of forensic science to solve criminal cases. When the original *CSI* reached the end of its run in 2015, CBS tried to extend the franchise with *CSI: Cyber*, but the new program—which emphasized internet-related investigations instead of forensic science—lasted only two seasons.

Along with these spinoffs, the network developed other programs featuring forensic science. In 2002, it launched *Without a Trace*, a show about FBI investigations of missing persons cases. A year later, two more forensic crime dramas debuted on CBS: *Cold Case*, a series about a criminal investigation unit using new technology to solve old cases, and *NCIS*, a series about agents working for the Naval Criminal Investigation Service. *Without a Trace* and *Cold Case* each aired for seven seasons. As for *NCIS*, it surpassed even *CSI*'s longevity when it entered its 16th season in the fall of 2018 (it was still in production as of 2021). It also exceeded *CSI*'s ratings success, spending a full decade in the top ten of the Nielsen ratings, with one season at *the* most-watched prime-time program. Though most of the main characters on *NCIS* are special agents, the show features one of the most prominent fictional scientists on television: Abby Sciuto, a cheerful goth with forensic skills in fields ranging from chemistry to computer science.

Meanwhile, other television networks followed suit with their own forensic-themed dramas. The most notable among these is *Bones*, a FOX series about a forensic anthropologist named Temperance Brennan (based on real-life scientist and author Kathy Reichs) and her team at the fictional Jefferson Institute in Washington, DC. The program premiered in 2005 and aired for 12 seasons. Brennan, as portrayed by Emily Deschanel, is a particularly interesting case of a prime-time scientist. Nicknamed "Bones" by her FBI partner, she fits some of the stereotypes entertainment media have historically used to portray men in science: she's socially clueless (one of her catchphrases is "I don't know what that means"), brusque, literal-minded, and more comfortable in her lab than elsewhere (though the program never identifies her as having an autism spectrum disorder, both Deschanel and show creator Hart Hansen have suggested the possibility).[27] She's also brilliant and admirable—as are her scientific colleagues, including Angela Montenegro and Camille Saroyan.

On cable television, multiple documentary-style programs have focused on forensic science in criminal investigations. Foremost among these is *Forensic Files*, which premiered as *Medical Detectives* on TLC in 1996 (shortly after the high-profile O. J. Simpson trial, which prominently featured DNA evidence) but underwent a name change in 2000 when it moved to Court TV. The series ran for 14 seasons. Other long-running reality television programs about forensic science have included the Discovery Channel's *The New Detectives* (1996–2004), which was developed by the creator of *CSI*, as well as A&E's *Cold Case Files* (1999–2017) and *The First 48* (2004–present).

The spinoffs, clones, and reality-based counterparts of *CSI* may have reinforced some of its key messages about forensic science: namely, that the forms of evidence they portray are highly reliable; that forensic approaches such as DNA testing, analysis of trace evidence, and analysis of bones or skeletons are commonly used in criminal investigations; and that forensic scientists are attractive heroes who live exciting lives as they help society by solving crimes.[28] Popular media accounts have even suggested new variations on the *CSI* effect linked to these shows. For example, a 2017 CNN story speculated that *Bones* might have produced a "Jeffersonian effect" on interest in forensic anthropology, particularly among young women.[29]

Studying the *CSI* Effect

So, can forensic crime television shows sway audience members? The answer may depend in part on the type of effect in question. To help sort through the possibilities, Simon Cole and Rachel Dioso-Villa assembled a rogue's gallery of potential *CSI* effects, including the "prosecutor's effect," the "defendant's effect," the "police chief's effect," the "educator's effect," and the "producer's effect."[30] Much of the research on the *CSI* effect(s) has focused on the first two suspects. The prosecutor's effect revolves around the possibility that crime television viewing will lead jurors to expect forensic evidence from the prosecution—and, thus, to reject cases that lack such evidence, no matter how strong. The defendant's effect, in turn, revolves around the possibility that crime television viewing will lead jurors to accept any forensic evidence the prosecution presents, no matter how flawed. Either sort of effect would imply potential distortions in the criminal justice process, with troubling implications.

Yet studies have yielded mixed results on whether watching crime television shapes how potential jurors perceive forensic science in trial settings.[31] In the "yes" (or at least "maybe") column, N. J. Schweitzer and Michael Saks found that *CSI* viewers were less likely than non-viewers to favor conviction in a simulated trial.[32] Likewise, Deborah Baskins and Ira Sommers concluded that potential jurors who watched high levels of crime television were more likely to say they would acquit in the absence of scientific evidence.[33] In the "no" column, Donald Shelton and his colleagues found that *CSI* viewers called for jury duty didn't differ from non-viewers on whether they saw forensic evidence as necessary for a guilty verdict.[34] Similarly, the *CSI* viewers Kimberlianne Podlas surveyed were no more likely than non-viewers to acquit based on "*CSI* factors."[35] As the quip-loving Gil Grissom might say, there's reasonable doubt when it comes to these sorts of *CSI* effects.

The evidence is also murky when it comes to the police chief's effect—that is, the potential for crime television viewing to influence how criminals behave. For example, does watching *CSI* lead perpetrators to wear gloves and bleach the scenes of their crimes so that they won't leave behind forensic evidence? To test this possibility, Andreas Baranowski and his colleagues conducted a creative experiment in which they asked two sets of participants—a group of crime television fans and a group of non-viewers—to commit a mock crime and clean a mock crime scene.[36] The fans left behind just as much forensic evidence as the non-viewers, suggesting that their media habits hadn't taught them any practical lessons about avoiding detection by criminal investigators.

By contrast, the popularity of forensic science on television does seem to have produced an educator's effect on interest in learning how to become a forensic scientist. The number of university programs offering forensic science degrees boomed after *CSI*'s debut, as did the number of students pursuing such degrees.[37] Furthermore, Sara McManus found that students' media habits—including their levels of forensic crime television viewing and forensic crime novel reading—were linked to their career interests in forensic science.[38] However, media messages may have given some students a skewed image of real-world forensic science, leaving them disillusioned with the field after learning more about it.

Last, but by no means least, *CSI* and shows like it may exert a producer's effect on how much audience members understand forensic evidence—or at least how much they think they understand it—and how much faith they place in it. For example, the *CSI* viewers Schweitzer and Saks surveyed were more likely than non-viewers to say they understood forensic science. Similarly, Baskins and Somers found a link between how much crime television their respondents watched and how reliable they perceived forensic evidence as being.

Beyond the *CSI* Effect

Two years after our trip to Las Vegas, we conducted our own tests of three potential *CSI* effects: the prosecutor's effect (on a willingness to acquit in the absence of forensic evidence), the defendant's effect (on a willingness to convict in the presence of forensic evidence), and the producer's effect (on a self-perceived understanding of and faith in forensic evidence). Building on the results of our content analysis, we focused on how crime television viewing might shape perceptions of one particularly prominent type of forensic evidence: DNA evidence.

But we didn't stop there, given that the forensic crime television genre isn't the only source of media messages about forensic science. Instead, we examined whether perceptions of DNA can reflect two other forms of media use, as well: overall television viewing and news media consumption. In terms of the former, we suspected two possibilities. Drawing on evidence that television viewing can *displace* opportunities to learn about science, we tested whether people who spent more time watching television would be less likely to say they understood DNA.[39] In light of findings that television portrayals can *cultivate* positive perceptions of science and scientists, we also tested whether higher levels of overall television viewing went hand in hand with greater faith in DNA evidence.[40]

As for news media use, we speculated that reading newspapers and watching local television news would be linked to understanding and trusting DNA evidence. Many news outlets devote extensive coverage to criminal investigations and trials.[41] Thus, people who follow these outlets may encounter a regular stream of news messages about forensic science. Furthermore, most news media coverage of science and technology tends to be positive—a pattern that presumably extends to forensic science and could help reinforce belief in the reliability of DNA evidence.[42]

To test these possibilities, we conducted a telephone survey in 2007, when *CSI, CSI: Miami, CSI: New York, Cold Case, NCIS, Bones,* and *Forensic Files* were all on the air.[43] We asked our respondents (908 residents of the Milwaukee area) one set of questions about their media habits and another set of questions about their perceptions of DNA, including their self-reported understanding of DNA, their views on the reliability of DNA evidence, their support for a national DNA databank, and their responses to the following hypothetical scenarios:

- Suppose that you were on a jury in a murder case, and the prosecution presented DNA evidence that linked the defendant to the crime. Would this make you much more likely to vote to convict, somewhat more likely, a little more likely, or no more likely to vote to convict?
- Suppose that you were on a jury in a murder case, and the prosecution DID NOT present DNA evidence that linked the defendant to the crime. Would this make you much more likely to vote to acquit, somewhat more likely, a little more likely, or no more likely to vote to acquit?

Contrary to some popular accounts of the *CSI* effect, our analysis didn't yield any "hits" when it came to matching crime television viewing habits and beliefs about how much weight DNA evidence should receive in a trial setting. Compared with respondents who seldom or never watched *CSI* and other forensic-themed television shows, those who frequently watched such shows were no more likely to say they would convict in the presence of DNA evidence. Nor were they more likely to say they would acquit in the absence of such evidence. In short, we found neither a prosecutor's effect nor a defendant's effect

On the other hand, we did find signs of a producer's effect on understanding and trusting DNA evidence (Figure 8.1). Among respondents who seldom or never watched forensic-themed television shows, 49% said they had a "clear understanding" of what DNA means, 21% said that DNA evidence is "completely" reliable (most of the rest said it was "very" reliable), and 54% supported a national databank of DNA from criminals. All three percentages were higher among respondents who frequently watched forensic-themed television shows: 62% for understanding what DNA means, 34% for believing DNA evidence is

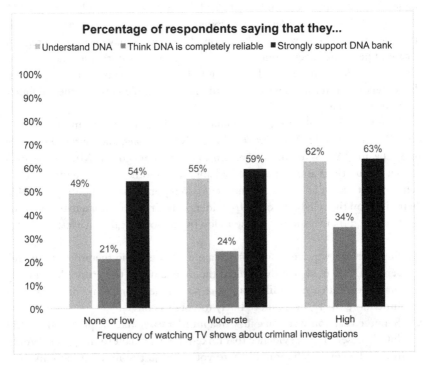

FIGURE 8.1 Perceptions of DNA, by crime TV watching (Institute for Survey and Policy Research at the University of Wisconsin-Milwaukee, 2007)

completely reliable, and 63% for supporting a DNA databank. Moreover, each of these patterns remained clear even when we statistically controlled for respondents' background characteristics.

Overall television viewing was also linked to perceptions of DNA evidence—though not always in the same ways as forensic crime television viewing (Figure 8.2). In keeping with the logic that time spent watching television can displace opportunities to learn about science, the respondents who watched more than three hours of television a day reported less understanding of DNA (51%) than the ones who watched little or no television (58%). At the same time, overall television viewing may cultivate faith in DNA evidence. Compared to respondents who watched little or no television, those who watched the most television were 14 points more likely to see DNA evidence as completely reliable and 10 points more likely to support a DNA databank.

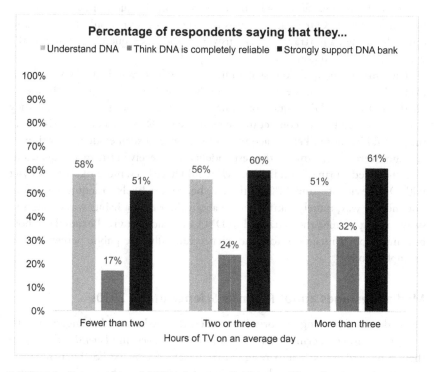

FIGURE 8.2 Perceptions of DNA, by overall TV watching (Institute for Survey and Policy Research at the University of Wisconsin-Milwaukee, 2007)

When we looked at whether perceptions of DNA reflected newspaper reading and local television news viewing, we found several clear patterns. Compared with non-readers, regular newspaper readers were 16 points likelier to say they understood what DNA means and 12 points likelier to support a DNA databank. In addition, they were 18 points likelier to support increased spending on DNA testing at the Wisconsin state crime lab—a proposal that had emerged as a political campaign issue after negative publicity surrounding a backlog in processing DNA samples. As for local television news viewing, it was linked to support for a DNA databank (62% for regular viewers versus 48% for non-viewers) but not to any other views about DNA.

Our survey also included an experiment that tested whether *priming* respondents to think about media portrayals would shape how they thought about forensic evidence. Priming theory suggests that highlighting specific pieces of information can alter how people form their decisions.[44] To capture this phenomenon, we randomly assigned each respondent to one of two versions of the survey: a version in which we asked the media questions *before* the questions about DNA, and a version in which we asked the media questions *after* the questions about DNA. Our logic here was that the first group would be primed to think about media portrayals when answering the DNA questions, whereas the second group would not.

The order of the questions shaped how respondents weighed DNA evidence in each hypothetical murder trial scenario. Among those who received the media questions after the DNA questions (the "no prime" condition), 85% said they would be more likely to convict in the presence of DNA evidence, and 34% said they would be more likely to acquit in the absence of such evidence. Both percentages were higher among the respondents who received the media questions first (the "media prime" condition): 91% would have been more likely to convict with DNA evidence, and 42% would have been more likely to acquit without it. Put another way, merely asking people about their media habits was enough to sway the importance they attached to DNA evidence—a vivid example of how priming popular images of forensic science can influence public perceptions of its implications.

Media Messages about Forensic Science in the 2010s

In the decade following our original studies, the media world changed in some ways. Many forensic crime shows, including *CSI*, *Bones*, and *Forensic Files*, ended their runs, and Abby Sciuto resigned from her post on *NCIS*. Even so, messages about forensic science have remained a fixture of the media landscape. Television networks may no longer be making new episodes about Gil Grissom solving crimes using entomology or Temperance Brennan identifying the bones of murder victims, but their work lives on in syndication and on streaming video services. Meanwhile, new characters—including Kasie Hines (Abby Sciuto's replacement on *NCIS*) and Sebastian Lund (from the spinoff *NCIS: New Orleans*)—have joined the ranks of prime-time forensic scientists. Similarly, news outlets have continued covering forensic science in the context of criminal cases, from the trial of Casey Anthony (in which Florida prosecutors failed to sway jurors with evidence obtained through controversial forensic techniques) to the murder of Hae Min Lee (the focus of *Serial*, one of a growing number of podcasts devoted to "true crime" along with broader issues in criminal justice).[45]

A particularly striking discussion of forensic science appeared on the October 1, 2017 episode of *Last Week Tonight with John Oliver*, HBO's satirical television news program. "This story is about how we increasingly solve crimes using forensic evidence," the show's host explains at the beginning of the segment. "It's that thing that is just a staple of TV crime shows." After playing clips from several shows, including *CSI*, *CSI: Miami*, and *Bones*, and making a joke about a monkey bite mark analysis portrayed in a *Law & Order* clip ("That last one was presumably from the one of the crossover episodes where the team from *Law & Order* worked a case with the cast of *Monkey Law & Monkey Order*"), he notes prosecutors' concerns about "the so-called *CSI* effect"—specifically, that television shows will lead jurors to expect forensic evidence in the courtroom.

From there, however, Oliver goes on to discuss how the National Academy of Sciences and a presidential council found that the forensic techniques portrayed on television vary substantially in how reliable they are in the real world. "It's not that all forensic science is bad, because it's not," he says, "But too often, its reliability is dangerously overstated." By way of example, he describes a case in which a flawed analysis of hairs found at a crime scene contributed to a wrongful murder conviction (subsequent DNA analysis showed that one of the hairs originally identified as belonging to the suspect came from a dog). He also highlights weaknesses in techniques used to analyze blood patterns, footwear, firearms, and bite marks.

As for fingerprints and DNA, Oliver calls them reliable but says that they are "by no means infallible." He then describes the case of the 2004 Madrid train bombings, in which a fingerprint match led to the arrest of a US citizen who had "never even been to Spain in his life." The host even points out that DNA testing, which he describes as "the gold standard in forensic science for a reason," can vary in its reliability, particularly when based on samples drawn from a messy crime scene. The segment concludes with a skit, designed to "educate potential jurors," that mocks forensic science as presented on shows such as *CSI*:

FORENSIC SCIENTIST 1: Chief, the hair matches the victim's wife. Case closed.
CHIEF: Slow down. Microscopic hair comparison is bullshit science.
FORENSIC SCIENTIST 2: Chief, I ran a mitochondrial DNA analysis on those hairs ...

FORENSIC SCIENTIST 1: The wife did it, right? Case closed.

FORENSIC SCIENTIST 2: Actually, there were five hairs. Three were from a coconut ... one was from a Cabbage Patch Kid, and the remaining one was from ... this golden retriever.

In short, Oliver argues that audience members should view media portrayals of forensic science with a healthy skepticism and draw distinctions between different forms of forensic science in terms of their real-life reliability.

Oliver's discussion raises two questions. First, have any links between people's media habits and their views of forensic science carried over to the 2010s? Second, do these patterns hold across different forms of forensic evidence, including ones with lower real-world reliability (such as hair and other trace materials) as well as ones with higher reliability (such as DNA)?

A New Look at Media Habits and Perceptions of Forensic Science

In 2016, we revisited the topic by conducting a new survey of the public.[46] This time, we asked 855 US residents how reliable they thought four forms of evidence were "for criminal investigations and trials": DNA evidence; fingerprint evidence; forensic examination of skeletons or teeth; and trace evidence such as fibers, hair, soil, wood, and gunshot residue. Their answers show that members of the public do draw distinctions between different forms of forensic evidence. Like the scientific experts John Oliver cites, our respondents tended to trust DNA the most and trace evidence the least. More than three-fourths of them (77%) gave DNA evidence the highest possible rating ("very reliable"). A smaller majority gave the same rating to fingerprints (57%) and forensic anthropology (56%), while fewer than half (44%) gave this rating to trace evidence. At the same time, respondents who saw any one form of forensic evidence as reliable tended to see the others as reliable, too.

When it came to entertainment television, more frequent exposure went hand in hand with greater faith in the reliability of forensic evidence (Figure 8.3). Compared with respondents who watched little or no entertainment television (less than two hours a day), those who watched the most (three-plus hours a day) were more likely to rate fingerprint evidence (60% versus 53%), analysis of bones or skeletons (67% versus 56%), and trace evidence (55% versus 41%) as very reliable. Interestingly, the one exception was DNA evidence: the six-point difference here (82% versus 76%) was not statistically meaningful.

Television news consumption also helped explain who saw forensic evidence as reliable (Figure 8.4). Here, higher viewing levels were linked to greater faith in DNA evidence: 86% of respondents who watched three-plus hours of television

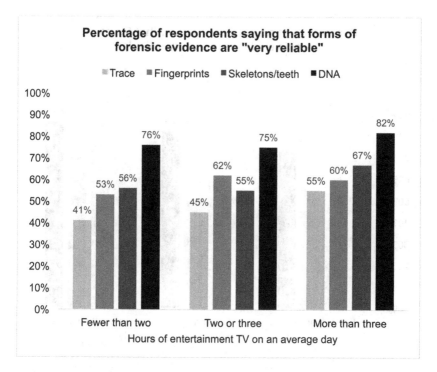

FIGURE 8.3 Perceptions of forensic science, by entertainment TV watching (Cooperative Congressional Election Survey, 2016)

news saw it as very reliable, whereas 73% of those who watched two hours or less said the same. Similar patterns emerged for seeing forensic anthropology (70% versus 55%), trace evidence (59% versus 41%), and fingerprint evidence (69% versus 54%) as very reliable.

The upshot here is that perceptions of forensic evidence as reliable reflected both entertainment television viewing and television news viewing. Indeed, these patterns were particularly clear when we looked at combined ratings for all four forms of evidence—even after we took other factors, including respondents' demographics, into account. Sixteen years after the first episode of *CSI* aired, media habits and public faith in forensic science remain intertwined.

Media Messages and Scientific Credibility

Despite all the talk about the potential for *CSI* and the like to shape what happens in courtrooms, the clues we've assembled here tell a different sort of story. As it turns out, messages about forensic science in entertainment crime television appear to matter less for how jurors make judgments about trials (or for how

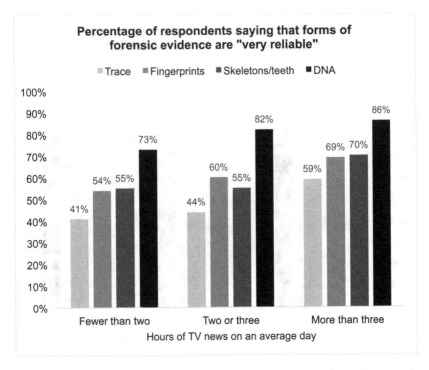

FIGURE 8.4 Perceptions of forensic science, by TV news watching (Cooperative Congressional Election Survey, 2016)

criminals behave) than for how audience members form their understandings of forensic science. Although priming media messages about forensic science in jurors' minds may help sway their verdicts, the clearest evidence for a *CSI* effect comes when we look at the broader public image of the field. People who watch forensic crime television place higher levels of faith in the reliability of forensic evidence, as do people who watch lots of entertainment television in general. Furthermore, prime-time portrayals of forensic science seem to have boosted the profession's appeal to students.

Yet entertainment crime television's depictions aren't the only media messages that can shape public perceptions of forensic science. For example, the people who watch the most television overall are particularly unlikely to think they understand DNA but particularly likely to trust DNA evidence. Meanwhile, news media habits help explain audience members' views on the credibility of specific forensic approaches, from DNA testing to analysis of trace evidence, along with public support for maintaining DNA databanks and spending money on crime labs. Focusing too narrowly on *CSI* and similar shows means missing the full picture of how media messages can help establish a mantle of credibility for forensic science.

Taken together, these effects could help shape viewers' own choices about what careers to pursue along with society's choices about how to fund and use forensic techniques in the criminal justice system and beyond. Our findings here also illustrate how seemingly realistic depictions of forensic science can bolster faith in its credibility, even when those portrayals exaggerate how swiftly the field works, how authoritative it is, or how glamorous it looks.

The same pattern could play out in other fields, as well. Messages in both entertainment media and news media can highlight—or challenge—scientific authority in a wide range of disciplines. For example, watching NASA scientists rescue a stranded astronaut in the 2015 film *The Martian* could bolster public faith in astronomy and astronomers, as could following news reports in 2019 about the first-ever photograph of a black hole. Similarly, portrayals of genetic engineering in news stories about real-world CRISPR applications and television dramas such as the cloning-themed *Orphan Black* could foster trust in, or doubts about, the technology.

The power of media messages to foster or undermine scientific authority may even extend beyond the boundaries of conventional science to "fringe" research on extra-sensory perception (ESP), unidentified flying objects (UFOs), ghosts, or "cryptids" such as Bigfoot and the Loch Ness monster. Hollywood has long been fascinated with these subjects, featuring them in an endless string of fictional movies and television shows—including more than a few that depict scientifically trained paranormal investigators. Likewise, many reality television programs and news reports have featured real-world paranormal investigators who stake claim to the language and methods of science to create an aura of credibility. So, do media messages influence perceptions of whether these investigators are scientific and whether the phenomena they study are real?

Notes

1 Pallotta, Frank, and Brian Stelter, "*CSI* to end after 15 seasons with a two-hour series sendoff," CNN Money, May 13, 2015, https://money.cnn.com/2015/05/12/media/csi-canceled/index.html.
2 Gerbner, George, Larry Gross, Michael Morgan, and Nancy Signorielli, "Living with television: The dynamics of the cultivation process," in *Perspectives on media effects*, eds. Jennings Bryant and Dolf Zillmann, Lawrence Erlbaum Associates, 1986: 17–40.
3 Kluger, Jeffrey, "How science solves crime," *TIME*, October 21, 2002: 36.
4 Cole, Simon A., and Rachel Dioso-Villa, "Investigating the 'CSI effect' effect: Media and litigation crisis in criminal law," *Stanford Law Review* 61, no. 6 (2009): 1335–1374.
5 Hill, Michael A., "Long before there was *CSI*, there was *Quincy*," *Chicago Tribune*, December 4, 2003, www.chicagotribune.com/news/ct-xpm-2003-12-04-0312040105-story.html.

6 Mann, Michael, "The *CSI* effect: Better jurors through television and science," *Buffalo Public Interest Law Journal* 24 (2005): 211–237; Turow, Joseph, *Playing doctor: Television, storytelling, and medical power*, University of Michigan Press, 2010.

7 Podlas, Kimberlianne, "'The *CSI* effect': Exposing the media myth," *Fordham Intellectual Property, Media and Entertainment Law Journal* 16, no. 2 (2005): 429–461.

8 Smith, Steven M., Veronica Stinson, and Marc W. Patry, "Fact of fiction: The myth and reality of the *CSI* effect," *Court Review: The Journal of the American Judges Association* 47 (2011): 4–7.

9 Cavender, Gray, and Sarah K. Deutsch, "*CSI* and moral authority: The police and science," *Crime, Media, Culture* 3, no. 1 (2007): 67–81; Deutsch, Sarah Keturah, and Gray Cavender, "*CSI* and forensic realism," *Journal of Criminal Justice and Popular Culture* 15, no. 1 (2008): 34–53; see also Tait, Sue, "Autoptic vision and the necrophilic imaginary in *CSI*," *International Journal of Cultural Studies* 9, no. 1 (2006): 45–62.

10 Kirby, David A., "Scientists on the set: Science consultants and the communication of science in visual fiction," *Public Understanding of Science* 12, no. 3 (2003): 261–278.

11 Kirby, David A., "Science consultants, fictional films, and scientific practice," *Social Studies of Science* 33, no. 2 (2003): 231–268.

12 Lieberman, Joel D., Courtney A. Carrell, Terance D. Miethe, and Daniel A. Krauss, "Gold versus platinum: Do jurors recognize the superiority and limitations of DNA evidence compared to other types of forensic evidence?" *Psychology, Public Policy, and Law* 14, no. 1 (2008): 27.

13 Bates, Benjamin R., "Public culture and public understanding of genetics: A focus group study," *Public Understanding of Science* 14, no. 1 (2005): 47–65; Condit, Celeste M., "How the public understands genetics: Non-deterministic and non-discriminatory interpretations of the 'blueprint' metaphor," *Public Understanding of Science* 8, no. 3 (1999): 169–180.

14 Federal Bureau of Investigation, "CODIS-NDIS statistics," Dec. 31, 2020, www.fbi.gov/ services/laboratory/biometric-analysis/codis/ndis-statistics.

15 Innocence Project, "DNA exonerations in the United States," Dec. 31, 2020, www. innocenceproject.org/dna-exonerations-in-the-united-states.

16 Ley, Barbara L., Natalie Jankowski, and Paul R. Brewer, "Investigating *CSI*: Portrayals of DNA testing on a forensic crime show and their potential effects," *Public Understanding of Science* 21, no. 1 (2012): 51–67.

17 *CSI: Crime Scene Investigation*, "Lady Heather's Box" (Season 3, Episode 15).

18 *CSI: Crime Scene Investigation*, "Pirates of the Third Reich" (Season 6, Episode 15).

19 *CSI: Crime Scene Investigation*, "Felonious Monk" (Season 2, Episode 17).

20 *CSI: Crime Scene Investigation*, "Invisible Evidence" (Season 4, Episode 7).

21 *CSI: Crime Scene Investigation*, "Table Stakes" (Season 1, Episode 15); *CSI: Crime Scene Investigation*, "Gentle, Gentle" (Season 1, Episode 19).

22 *CSI*, "Table Stakes."

23 *CSI: Crime Scene Investigation*, "Sex, Lies, and Larvae" (Season 1, Episode 10); *CSI: Crime Scene Investigation*, "Cool Change" (Season 1, Episode 2).

24 *CSI: Crime Scene Investigation*, "$35K O.B.O" (Season 1, Episode 18).

25 Nelkin, Dorothy, and M. Susan Lindee, *The DNA mystique: The gene as a cultural icon*, University of Michigan Press, 2010.

26 Bandura, Albert, "Human agency in social cognitive theory," *American Psychologist* 44, no. 9 (1989): 1175–1184; Bandura, Albert, "Social cognitive theory of mass communication," *Media Psychology* 3, no. 3 (2001): 265–299; Long, Marilee, and Jocelyn Steinke, "The thrill of everyday science: Images of science and scientists on children's educational science programmes in the United States," *Public Understanding of Science* 5, no. 2 (1996): 101–120; Steinke, Jocelyn, Maria Knight Lapinski, Nikki Crocker, Aletta Zietsman-Thomas, Yaschica Williams, Stephanie Higdon Evergreen, and Sarvani Kuchibhotla, "Assessing media influences on middle school-aged children's perceptions of women in

science using the Draw-A-Scientist Test (DAST)," *Science Communication* 29, no. 1 (2007): 35–64.

27 Malia, Jennifer, "I'm an autistic woman, and Bones is the only character like me on TV," *Glamour*, March 28, 2017, www.glamour.com/story/bones-series-finale.

28 Rhineberger-Dunn, Gayle, Steven J. Briggs, and Nicole E. Rader, "The *CSI* effect, DNA discourse, and popular crime dramas," *Social Science Quarterly* 98, no. 2 (2017): 532–547.

29 Gonzalez, Sandra, "How *Bones* bred a new generation of female scientists," CNN Entertainment, March 27, 2017, www.cnn.com/2017/03/27/entertainment/bones-tv-show-women-stem/index.html.

30 Cole, and Dioso-Villa, "Investigating the '*CSI* effect' effect," 1343–1344.

31 Cole, Simon A., and Glenn Porter, "The *CSI* effect," in *Routledge International Handbook of Forensic Intelligence and Criminology*, eds. Quentin Rossy, David Décary-Hétu, Olivier Delémont, and Massimiliano Mulone, Routledge, 2017: 112–124.

32 Schweitzer, Nicholas J., and Michael J. Saks, "The *CSI* effect: Popular fiction about forensic science affects the public's expectations about real forensic science," *Jurimetrics* 47, no. 3 (2007): 357–364.

33 Baskin, Deborah R., and Ira B. Sommers, "Crime-show-viewing habits and public attitudes toward forensic evidence: The '*CSI* effect' revisited," *Justice System Journal* 31, no. 1 (2010): 97–113.

34 Shelton, Donald E., Young S. Kim, and Gregg Barak, "A study of juror expectations and demands concerning scientific evidence: Does the '*CSI* effect' exist?" *Vanderbilt Journal of Entertainment & Technology Law* 9 (2006): 331–368; Shelton, Donald E., Young S. Kim, and Gregg Barak, "An indirect-effects model of mediated adjudication: The *CSI* myth, the tech effect, and metropolitan jurors' expectations for scientific evidence," *Vanderbilt Journal of Enterntainment & Technology Law* 12, no. 1 (2009): 1–43.

35 Podlas, "'The *CSI* effect'"; Podlas, Kimberlianne, "The *CSI* effect and other forensic fictions," *Loyola of Los Angeles Entertainment Law Review* 27, no. 2 (2006): 87–125.

36 Baranowski, Andreas M., Anne Burkhardt, Elisabeth Czernik, and Heiko Hecht, "The *CSI*-education effect: Do potential criminals benefit from forensic TV series?" *International Journal of Law, Crime and Justice* 52 (2018): 86–97.

37 Bergslien, Elisa, "Teaching to avoid the '*CSI* effect': Keeping the science in forensic science," *Journal of Chemical Education* 83, no. 5 (2006): 690–691; Jackson, Glen Paul, "The status of forensic science degree programs in the United States," *Forensic Science Policy and Management* 1, no. 1 (2009): 2–9; McManus, Sarah E., "Influence of the *CSI* effect on education and mass media," MA thesis, University of Tennessee, 2010.

38 McManus, "Influence of the *CSI* effect."

39 Dudo, Anthony, Dominique Brossard, James Shanahan, Dietram A. Scheufele, Michael Morgan, and Nancy Signorielli, "Science on television in the 21st century: Recent trends in portrayals and their contributions to public attitudes toward science," *Communication Research* 38, no. 6 (2011): 754–777.

40 Nisbet, Matthew C., Dietram A. Scheufele, James Shanahan, Patricia Moy, Dominique Brossard, and Bruce V. Lewenstein, "Knowledge, reservations, or promise? A media effects model for public perceptions of science and technology," *Communication Research* 29, no. 5 (2002): 584–608. See also Chapter 3.

41 Kerbel, Matthew Robert, *If it bleeds, it leads: An anatomy of television news*, Routledge, 2018.

42 Besley, John C., and James Shanahan, "Media attention and exposure in relation to support for agricultural biotechnology," *Science Communication* 26, no. 4 (2005): 347–367; Brossard, Dominique, and James Shanahan, "Do citizens want to have their say? Media, agricultural biotechnology, and authoritarian views of democratic processes in science," *Mass Communication and Society* 6, no. 3 (2003): 291–312; Lee, Chul-joo, and Dietram A. Scheufele, "The influence of knowledge and deference towar-d scientific

authority: A media effects model for public attitudes toward nanotechnology," *Journalism & Mass Communication Quarterly* 83, no. 4 (2006): 819–834; Liu, Hui, and Susanna Priest, "Understanding public support for stem cell research: media communication, interpersonal communication and trust in key actors," *Public Understanding of Science* 18, no. 6 (2009): 704–718.

43 Brewer, Paul R., and Barbara L. Ley, "Media use and public perceptions of DNA evidence," *Science Communication* 32, no. 1 (2010): 93–117.

44 Iyengar, Shanto, and Donald R. Kinder, *News that matters: Television and American opinion*, University of Chicago Press, 2010.

45 Grinberg, Emanuella, "Flawed forensic evidence explains Casey Anthony acquittal, experts say," CNN, July 18, 2011, http://edition.cnn.com/2011/CRIME/07/15/casey.anthony.forensic.evidence; Prudente, Tim, "After *Serial* podcast, prosecutors tested DNA evidence in Adnan Syed case. Here's what they found," *Baltimore Sun*, March 28, 2019, www.baltimoresun.com/news/maryland/crime/bs-md-ci-syed-dna-evidence-20190328-story.html.

46 See the Appendix for details.

9

FRINGE SCIENCE

Dr. Peter Venkman: Back off, man. I'm a scientist.
——*Ghostbusters* (1984)

Grant Wilson: The K-II Meter measures magnetic fields, and it's been specially calibrated for paranormal investigators. The theory behind the K-II is that if there is a spirit in the area, the K-II meter will pick up its magnetic field.
——*Ghost Hunters* (2007; Season 3, Episode 12: The Manson Murders)

When we were growing up, one of us (specifically, Paul) wanted to be a paranormal investigator just like the heroes of the 1984 movie *Ghostbusters*: Dr. Peter Venkman, Dr. Ray Stantz, Dr. Egon Spengler, and Winston Zeddemore. Three of these characters were former professors of parapsychology, and all of them were adept at wielding the team's ghost-hunting equipment—including their proton packs, which fired beams from "unlicensed nuclear accelerators" to capture "negatively charged ectoplasmic entities." The Ghostbusters cracked jokes and made weird science look cool as they defeated a slimy poltergeist, a giant marshmallow man, and an evil god named Gozer the Gozerian.

Along the way, the movie's heroes became pop culture icons. *Ghostbusters* was a box office hit, spawning a 1989 sequel, an animated television series, a 2016 reboot, and an additional sequel planned for 2021. It also inspired a legion of trick-or-treaters, including four fictional middle schoolers from the Netflix drama *Stranger Things* who dressed up as the Ghostbusters for Halloween in the show's second season.

DOI: 10.4324/9781003190721-9

Though *Ghostbusters* aimed for laughs, its dialogue was filled with scientific-sounding jargon. Of the characters, Stantz was especially fond of talking about "free floating, full torso, vaporous apparitions," "ectoplasmic residue," and so on. That's because Dan Ackroyd, who played him, drew on his own fascination with the paranormal in co-writing the movie's screenplay. For him, the topic was part of a family tradition: his great grandfather was a medium, his grandfather tried to contact spirits using radio waves, his father wrote a book about ghosts, and the actor himself claims to have encountered an "unseen presence."[1] By contrast, his screenwriting partner Harold Ramis—who played Spengler—was a skeptic.[2]

As Ackroyd's family history illustrates, real-life paranormal investigators have been searching for spooks for more than a century. Some of these supernatural sleuths have emphasized spiritualism and séances, but many modern ghost hunters describe themselves in a more scientific-sounding way.[3] For example, Jason Hawes founded TAPS (The Atlantic Paranormal Society) in 1990 to help "those experiencing paranormal activity by investigating claims in a professional and confidential manner … implementing the latest in paranormal research equipment and techniques."[4] He and his partner, Grant Wilson—both originally plumbers by trade—have spent three decades searching for signs of spectral presences, from "orbs" to "cold spots" to "EVPs" (electronic voice phenomena), using a variety of devices: infrared cameras, EMF (electromagnetic field) detectors, digital thermometers, and more.

Hawes and Wilson became television stars in 2004 with the launch of *Ghost Hunters*, a "reality" television show on the SyFy cable channel. The series ran for 12 seasons, with each episode focusing on one or more supposed hauntings. In some cases, the investigators from TAPS debunk claims of ghosts. In others, they conclude that their observations can't be explained away as normal phenomena.

At its peak, *Ghost Hunters* was SyFy's top-rated program, drawing 3 million viewers per episode.[5] The success of the series inspired a wave of spin-offs and copycat shows featuring investigators who track ghosts, UFOs (unidentified flying objects) and aliens, or "cryptids" (animals not recognized by mainstream science) such as Bigfoot, the Loch Ness Monster, and the Chupacabra. Meanwhile, skeptics dismissed the techniques used on these programs as "junk science." For example, the Committee for Skeptical Inquiry's Benjamin Radford pointed out that the investigators on *Ghost Hunters* possess "little or no training in science" and conduct research so flawed as to be "scientifically worthless."[6] Likewise, author and former magician Joe Nickell criticized their logical fallacies ("We don't know what caused such-and-such—a noise, say—so it must have been a ghost") and "pseudoscientific" use of equipment.[7]

Most scientists would undoubtedly agree with Radford and Nickell: as we write this, no one has managed to capture a ghost, alien, or Sasquatch. Nevertheless, a substantial proportion of Americans believe in paranormal phenomena. In 2005, the year after *Ghost Hunters* premiered, a national Gallup poll found that three in four respondents held at least one paranormal belief.[8] Fully 41% believed in ESP, or extra sensory perception; 37% believed that houses can be haunted; and 21% believed that people can communicate mentally with someone who has died. Similarly, more than a quarter (27%) of those polled by CBS News in 2014 thought it was somewhat or very likely that Bigfoot exists, and 40% of those polled by the same organization in 2017 believed in ghosts.[9] As for extraterrestrials, 33% of the respondents in a 2019 Gallup poll said that some UFOs have been alien spacecraft visiting Earth from other planets or galaxies.[10]

Why do so many people hold these beliefs, even in the face of the scientific community's skepticism about paranormal phenomena? One important point to remember is that people can filter what they hear about science—including "fringe science"—through their own worldviews.[11] Just as some conservatives reject climate science on ideological grounds and some evangelicals reject evolution on the basis of religious faith, believers in the paranormal may avoid, rationalize away, or outright reject scientific criticisms of the cases for alien visitors, Bigfoot, or haunted houses. Similarly, believers could filter scientific information through their own personal experiences.[12] For example, people who believe they've seen UFOs or experienced ESP may trust the evidence of their own senses—either the traditional five or a sixth—over the conclusions of the scientific community.

Another point to bear in mind is that interest and belief in the paranormal is, sociologically speaking, normal. Building on surveys of the public along with participant observation of haunted house investigations, Bigfoot hunts, and UFO conventions, a team of Baylor University researchers found that a wide variety of Americans believe in paranormal phenomena.[13] Some of these believers are socially unconventional, but many are not. Even between the two of us, Paul is a skeptic in the mold of Harold Ramis whereas Barbara, like Dan Ackroyd, has experienced potential instances of the paranormal and is more open to the possibility of such phenomena. Despite our differing viewpoints, each of us has been fascinated with the topic since childhood. Barbara grew up reading books about subjects ranging from the Bermuda Triangle to ESP, Paul has written science fiction about steampunk spirit-hunters, and both of us enjoy watching shows such as *Lost, Fringe,* and *Stranger Things.*

Speaking of popular images, some observers argue that media messages—including sensationalized news coverage, documentary-style shows, and television dramas—help fuel public faith in ghost hunting, ufology, cryptozoology, and other forms of fringe science. During the 1990s, skeptics such as Matthew Nisbet and Richard Dawkins accused *The X-Files,* a popular television drama, of undermining scientific literacy and promoting paranormal beliefs.[14] A decade later, fellow skeptics Benjamin Radford and Alison Smith leveled similar charges against *Ghost Hunters* and its ilk.[15]

Like all hypotheses, these claims about media influence on paranormal beliefs warrant careful investigation. Accordingly, we've assembled the data on the topic—some collected during the "paranormal boom" of the 1990s, and some gathered by us over the past ten years. Examining the evidence here also gives us a case study for exploring the broader question of how media messages can bolster or undermine claims to scientific authority for endeavors that lie outside the bounds of mainstream science. The results show that media messages presenting fringe researchers as scientific can promote faith in the paranormal. At the same time, messages debunking or poking fun at such researchers can undercut their credibility and reduce belief in phenomena such as UFOs, haunted houses, and ESP.

Paranormal Media in the 1990s

Media attention to, and public interest in, the paranormal has ebbed and flowed over the years, driven by shifting social values and cultural anxieties.[16] In the wake of World War I, newspapers ran articles hyping the claims of mediums along with the efforts of magician Harry Houdini and other skeptics to debunk them. During the early years of the Cold War, media attention shifted to UFOs following stories of "flying saucers" and the crash of a "flying disc" (ultimately reported by the US military to be a weather balloon) in Roswell, New Mexico. As the New Age movement emerged in the 1970s, Uri Geller drew fame for his claims of psychic spoon-bending powers—and ridicule when he failed to demonstrate those powers on an episode of *The Tonight Show* (the show's host, Johnny Carson, had enlisted the aid of magician James Randi to thwart any trickery on Geller's part).[17]

A pair of fictional FBI agents, Fox Mulder and Dana Scully, helped usher in yet another paranormal media boom when *The X-Files* debuted on FOX in 1993. Mulder is the show's advocate for the paranormal: his office sports a UFO poster that proclaims, "I WANT TO BELIEVE," and his familiarity with fringe lore helps the duo solve mysteries involving aliens, hauntings, cryptids, and other strange phenomena. By contrast, Scully is the program's resident skeptic: a medical doctor, she challenges Mulder's theories with scientific reasoning. In most cases, Mulder turns out to be right—as reflected in the show's mantra, "The truth is out there."

The success of *The X-Files* spurred FOX and its rival networks to launch other paranormal-themed shows, including the short-lived spinoff *The Lone Gunmen*, the teen-alien soap opera *Roswell*, and the psychic-sleuth drama *Medium*. Mulder and Scully's popularity no doubt also influenced FOX's decision in 1995 to air a "documentary" titled *Alien Autopsy: Fact or Fiction?* Though derided as an obvious hoax—and ultimately revealed to be fake—so many Americans tuned in that the network aired it twice more, drawing millions of viewers each time.[18]

Meanwhile, an array of news media outlets offered their own extraterrestrial-themed coverage. Even staid newspapers such as the *New York Times* and *Washington Post* ran hundreds of stories about UFOs and aliens in the mid-to-late 1990s.[19]

As the decade neared its end, ratings for *The X-Files* declined and the media's preoccupation with UFOs faded, only for another paranormal topic to take center stage: psychic abilities.[20] The leading figure in this shift was John Edward, whose television series *Crossing Over* premiered on the SyFy Channel in 2000. Edward, a self-proclaimed medium, told viewers he could use his connection with the spirit world ("the other side") to divine facts about the guests on his show. Though he became a media sensation, skeptics were not impressed.[21] As Joe Nickell explained, Edward was actually using a pair of old fortune-telling tricks—"cold reading" and "hot reading"—that Harry Houdini had debunked long before Edward was even born.[22]

Like the UFO hype, the media's fixation on speaking with the dead didn't last long: after the terrorist attacks on September 11, 2001 news coverage of psychic powers faded to less than half its peak in 1999–2000.[23] Edward himself sparked a backlash when he unveiled plans to "contact" 9/11 victims in a televised special; after a barrage of negative publicity, his producers shelved the episode.[24] A few years later, *Crossing Over* passed on into the realm of cancellation.

Media Effects on Paranormal Beliefs: Cultivation and Priming

During the height of this paranormal boom, Glenn Sparks and his colleagues conducted a series of studies testing whether media messages can *cultivate* public beliefs about fringy topics. Drawing on the idea that television is the leading mythmaker for modern society, they speculated that frequent portrayals of UFOs, psychic powers, and the like on television would encourage viewers to accept such phenomena as real.[25]

In 1994, Sparks and his team surveyed members of the public about their television habits and paranormal beliefs.[26] The researchers didn't find a clear link between overall television viewing and beliefs in phenomena as ghosts and UFOs, but they did find that watching paranormal-themed television went hand in hand with believing in such phenomena. When Sparks and Wes Miller conducted a larger study three years later, they found that overall television viewing was weakly connected to believing in a broad range of fringy phenomena (from ESP to astral projection to alien abductions) whereas paranormal television viewing was strongly tied to the same beliefs.[27] As an *X-Files* fan might put it, both studies suggest that media messages—particularly paranormal-themed television shows—can lead audience members to think more like Mulder and less like Scully.

Extending these findings, Matt Nisbet argues that media depictions of aliens and psychics can *prime* audience members to think of such portrayals when judging whether paranormal phenomena are real. "Individual opinion is heavily dependent on the types of considerations and examples about a topic that are available in short term memory," he explains.[28] "Which depictions of reality are more accessible in a person's mind than others is a function of the nature of media content and the media consumption habits of the individual." For instance, the 1990s-era hype surrounding paranormal topics may have filled many Americans' heads with easy-to-remember images of UFOs and psychics.

Looking at Gallup polls, Nisbet found that cycles in media attention dovetailed with shifts in public beliefs about the paranormal. The percentage of Americans who believed that extraterrestrials had visited the earth increased from 27% in 1990 to 33% in summer 2001—a period that encompassed the rise of *The X-Files*, the furor over *Alien Autopsy*, and a spate of UFO news stories. The figure then declined to 24% in 2005, by which time media attention to aliens had fallen to less than a third of its peak. Likewise, belief in communication with the dead rose from 22% in 1996 to 28% in summer 2001, when John Edward was basking in his pre-9/11 fame, before dropping to 23% in 2005 in tandem with fading media attention to psychics. As Nisbet points out, these trends are exactly what one would expect based on priming theory.

Paranormal Media through the 2010s

Though *The X-Files* and *Crossing Over* had both ceased production by the end of 2004, the premiere of *Ghost Hunters* that same year helped trigger the first paranormal media boom of the 21st century. When Jason Hawes and Grant Wilson's show became a hit, SyFy created a pair of spinoffs: *Ghost Hunters International* (which debuted in 2008) and *Ghost Hunters Academy* (2009). Other cable networks followed suit with their own ghost-themed programs, including A&E's *Paranormal State*, Animal Planet's *The Haunted*, Biography's *Celebrity Ghost Stories*, the Discovery Channel's *Ghost Lab*, and the Travel Channel's *Ghost Adventures*. The growing popularity of ghost-hunting also inspired a wave of news coverage: for example, the *Washington Post* ran a story about the D.C. Metro Area Ghost Watchers, and the *New York Times* profiled Twisted Dixie, a team of paranormal investigators in South Carolina.[29]

Along with this surge of attention to ghosts, the early 2000s saw a revival of interest in aliens and UFOs. The History Channel, a cable network formerly best known for its World War II documentaries, led the way in extraterrestrial-themed reality programming with *UFO Files* (which ran from 2004 to 2007) and *UFO Hunters* (2008–2009). In 2010, the network launched its most famous entry into the genre: *Ancient Aliens*, a documentary series about supposed human contacts with extraterrestrial civilizations. The program elicited derision from the scientific community; for example, science writer Riley Black described it as "a slimy and incomprehensible mixture of idle speculation and outright fabrications," and archaeologist Ken Feder called it "execrable bullshit."[30] Yet *Ancient Aliens* was not only a ratings success, with 14 seasons and counting, but also a pop culture phenomenon: the animated television show *South Park* parodied it, and one of its commentators (Giorgio A. Tsoukalos, also known as "the Aliens guy") became the subject of a famous internet meme.

The post-*Ghost Hunters* paranormal boom extended to cryptozoology, as well. Bigfoot, a hairy hominid purported to dwell in the Pacific Northwest, enjoyed a brief spell as a celebrity in the 1970s thanks to its out-of-focus appearances in pseudo-documentaries and its fights with the title character of *The Six Million Dollar Man* on prime-time television. In 2011, the Sasquatch became a television star again with the premiere of Animal Planet's *Finding Bigfoot*, a reality show following three cryptozoologists and one skeptic as they hunted their elusive quarry. They never found it, even after nine seasons of searching, but the show was a ratings success.

During all of this, the broadcast networks made numerous attempts to capture the public's imagination with paranormal-themed dramas. The most successful of them was *LOST*, which premiered on ABC in 2004 and drew high ratings throughout its six-year run. The show depicted many types of paranormal phenomena—including visions of the future and communication with the dead—in its story of airplane crash survivors stranded on a mysterious island. A flood of copycat shows followed, including *Flash Forward*, *Threshold*, and *The Event*. Most vanished quickly, but *LOST*'s creator, J.J. Abrams, developed another paranormal-themed series, *Fringe*, that ran for five seasons (2008–2013) to considerable critical acclaim. This show followed in the footsteps of *The X-Files* by portraying a team of heroes—a mad scientist, his son, and an FBI agent—who investigate a wide variety of paranormal cases.

Mulder and Scully themselves returned to television in 2016 with a new season of *The X-Files* on FOX. The show's revival drew respectable ratings but was ultimately overshadowed by another paranormal drama that debuted the same year: *Stranger Things*. Produced by the streaming video service Netflix, it chronicled the weird science afoot during the 1980s at a US Department of Energy Laboratory. One key character in *Stranger Things* is Eleven, a girl with psychic powers that include telekinesis and clairvoyance. With her help, a group of middle-schoolers, teenagers, and adults work to defeat both the evil scientist in charge of the lab and dangerous invaders from a shadowy alternate reality nicknamed "the Upside Down." The show was a record-breaking success for Netflix, drawing 64 million viewers during its third season.[31]

A New Look at Media Habits and Paranormal Beliefs

Our own investigation into how media messages shape paranormal beliefs started in 2011, when *Ghost Hunters* was peaking in popularity and copycat shows were proliferating on cable television. Surveying 520 members of the public, one of us (Paul) found that paranormal *reality* television viewing went hand in hand with believing in ghosts and haunted houses—as did overall television viewing.[32] Moreover, both paranormal reality television viewing and total television viewing were linked to believing that paranormal investigators are scientific. By contrast, the survey didn't find any link between watching paranormal *dramas* on television and believing in ghosts or haunted houses, or in seeing ghost hunters as scientific.

These findings made us curious about whether audience members respond in different ways to fictional shows about the paranormal, such as *The X-Files* and *Lost*, versus documentary-style shows about the same topic, such as *Ghost Hunters*, *Ancient Aliens*, and *Finding Bigfoot*. We also wanted to take a closer look at whether overall television viewing can cultivate paranormal beliefs among the broader public. After all, messages about the paranormal appear in a broad range of television programming, from news programs (which have covered UFO sightings and ESP research) to crime dramas such as *CSI* (where the show's heroes investigated the murder of a ghost hunter in one episode) to animated comedies such as *The Simpsons* (which once featured cameos by Mulder and Scully).[33]

In the fall of 2018, we surveyed a representative sample of 1,000 Americans.[34] To capture their media habits, we asked them how many hours of television they watched on an average day, along with whether they had watched six specific shows. Three of these were paranormal reality shows: *Ghost Hunters* (which 37% of the respondents had seen), *Ancient Aliens* (34%), and *Finding Bigfoot* (24%). Likewise, three of the shows were paranormal-themed dramas: *The X-Files* (56%), *Lost* (36%), and *Stranger Things* (35%). To measure respondents' views on the paranormal, we asked them whether they believed that "some people have ESP, or extrasensory perception," "houses can be haunted," "extra-terrestrial beings have visited earth at some point in time in the past," "people can communicate mentally with someone who has died," and "creatures such as Bigfoot or the Loch Ness Monster may exist."

In keeping with our earlier research, we found a strong link between paranormal reality television viewing and paranormal beliefs (Figure 9.1). Compared to respondents who had seen zero or one shows in the genre, those who had seen two or three were more likely to believe in fringy phenomena such as ESP (by 13 percentage points), haunted houses (17 points), extraterrestrial visitors (25 points), communication with the dead (15 points), and cryptids (17 points).

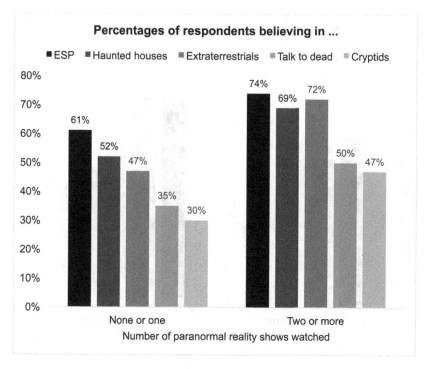

FIGURE 9.1 Beliefs about paranormal phenomena, by paranormal reality TV watching (Cooperative Congressional Election Survey, 2018)

Moreover, the relationship between paranormal reality television viewing and paranormal beliefs was clear even after we accounted for other media habits and background factors.

Of course, people who already believe in the paranormal may be particularly drawn to media messages about the topic. Still, it's striking that we found no link between paranormal *drama* viewing and paranormal beliefs once we accounted for paranormal reality television viewing and other factors. Watching shows such as *Ghost Hunters*, *Ancient Aliens*, and *Finding Bigfoot*—all of which use a documentary style and purport to be factual—went hand in hand with believing in haunted houses, extraterrestrial visitors, and the like, but the same was not true for watching self-evidently fictional shows such as *The X-Files*, *Lost*, and *Stranger Things*.

Looking beyond specific genres, we also found a link between total television viewing and belief in the paranormal (Figure 9.2). Compared to their no-or-low viewing peers, the respondents in our survey who watched 3 or more hours of television a day were more likely to believe in ESP (by

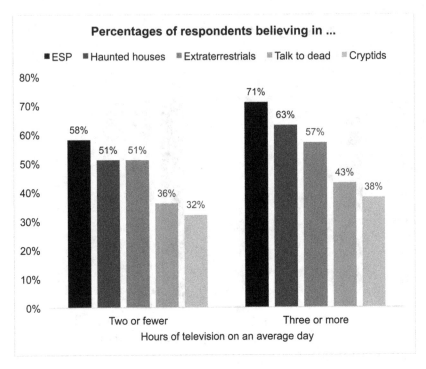

FIGURE 9.2 Beliefs about paranormal phenomena, by overall TV watching (Cooperative Congressional Election Survey, 2018)

13 points), haunted houses (2 points), extraterrestrial visitors (6 points), communication with the dead (7 points), and creatures such as Bigfoot and the Loch Ness Monster (6 points). These gaps weren't huge, but the overall relationship was statistically meaningful even when we considered the role of paranormal television viewing and a host of alternative suspects. We're skeptical that paranormal beliefs shape people's general viewing habits: why would believing in ghosts or UFOs make people watch more television? Maybe the Mulders among us are hunkering in front of their screens to hide from aliens or Sasquatches, but it seems more plausible that watching television cultivates paranormal beliefs.

Promoting Belief or Skepticism?

Thus far, we've focused on the broad connections between Americans' television habits and their beliefs about fringy phenomena. Yet media messengers present a variety of perspectives on the paranormal. Not only do they differ on whether ghosts, extraterrestrial visitors, and the like are real; they also differ in how they make their cases to audience members. For example, some arguments for the existence of hauntings rely on "spectral evidence" from mediums and seances, whereas other arguments—such as the ones presented by the investigators on *Ghost Hunters*—invoke science. Similarly, some skeptics—including Bill Nye and Neil deGrasse Tyson—have cited scientific evidence to debunk claims that paranormal phenomena are real, whereas other messengers—including late-night television hosts Johnny Carson and Stephen Colbert—have used humor to satirize such claims. So, what sorts of media messages are most effective at promoting belief in paranormal phenomena, and what sorts of messages work to foster skepticism about such phenomena?

Two decades ago, Glenn Sparks and his colleagues conducted a series of experiments to explore these topics. Their first study focused on *Beyond Reality*, a documentary-style television show about paranormal investigators.[35] The researchers randomly assigned each participant to one of four conditions: a group that watched an episode about astral projection (that is, the ability to send one's consciousness outside one's body to other places or planes of existence); a group that watched the same episode, but with a disclaimer that the program was fictitious; a group that watched the episode with a stronger disclaimer that the program was unscientific; or a group that didn't watch the episode. Watching the episode *without* a disclaimer led participants to express greater belief in the paranormal immediately afterward and even three weeks later. Meanwhile, watching the episode *with* a disclaimer led viewers to express greater skepticism about the paranormal. In short, adding a disclaimer reversed the impact of the message.

The next experiment conducted by Sparks and his colleagues offered further evidence that media messages can shape beliefs about the paranormal.[36] In this study, the participants watched a segment from *Unsolved Mysteries*, a documentary-style television program. Some watched a segment about sightings of flying saucers and space aliens, while others watched a segment about a non-paranormal mystery. Afterward, the participants in the first group were more likely to believe in UFOs.

A follow-up experiment tested the impact of quoting a scientist in a news story about the paranormal.[37] This time, the researchers gave participants different versions of a news magazine story about aliens abducting humans. A version that cited a scientist ("a Pulitzer Prize-winning psychiatrist at Harvard University") as endorsing a paranormal explanation bolstered belief in UFOs. Meanwhile, a version that presented claims of alien abductions without any scientific backing had no effect on viewers' beliefs.

The final experiment conducted by Sparks and his team found that the impact of media messages about the paranormal—in this case, about an alleged cover-up of a UFO crash near Roswell, New Mexico—can depend on whether they "balance" arguments from believers and skeptics.[38] The researchers showed one group of participants an edited version of a television news segment (from the CBS program *48 Hours*) that presented a paranormal explanation for the event and another group a version of the segment that featured a scientist debunking claims about UFOs. Compared to the participants who watched the one-sided version of the story, those who saw the two-sided version came away less likely to believe in UFOs.

Taken together, these studies show that media messages can boost or reduce belief in the paranormal, depending on how they portray the topic. The findings from the last study also piqued our interest in whether investigators like the ones on *Ghost Hunters* can enhance their credibility with the public by presenting themselves as scientific—as well as whether skeptics can undermine faith in fringy science by challenging such investigators' methods.

Framing Paranormal Investigators: The Trappings of Science

Media messengers often use the "trappings of science" to convey scientific authority.[39] As we've seen, many movies and television shows depict scientists in white lab coats, surrounded by equation-filled chalkboards, bubbling test tubes, or elaborate contraptions. Likewise, both fictional and real-life accounts of scientific work frequently quote experts who use complex terminology. For good reason: research shows that messages featuring the "look" and "language" of science can lead audience members to accept even wildly inaccurate claims.[40] Accordingly, paranormal investigators may be able to bolster their own powers of persuasion by *framing* themselves as scientific.

Inspired by *Ghost Hunters*, one of us (Paul again) designed an experiment to test whether media messages featuring science-y terminology and technology can influence audience members' views on the paranormal.[41] He crafted three different versions of a news story about a ghost-hunter (based on a real *Washington Post* profile): one that emphasized spiritualism and supernatural explanations, one that invoked scientific authority on behalf of paranormal claims, and one that invoked the same authority but also featured a scientific rebuttal. The first version discussed the investigator's childhood experiences with ghosts, his training as a medium, and his "openness to nonscientific methods." By contrast, the second version highlighted his "meticulous approach," quoted him using ghost-hunting jargon ("EVPs"), and described his fancy equipment (digital recorders, night-vision cameras, and EMF detectors). The third version was identical to the second except that it also pointed out the absence of "globally accepted guidelines" for paranormal investigations, compared such investigations to "old-time medicine shows," and quoted a skeptical psychology professor ("Despite all the trappings, it isn't science.").

Paul randomly assigned the respondents in his 2011 survey to read one of these three versions or a story on a non-paranormal topic. The version that highlighted supernatural accounts and spiritualist methods didn't sway audience members (Figure 9.3): the ones who read it weren't any more likely than the control group to believe in haunted houses (28% versus 40%), or to see paranormal investigators as scientific (40% versus 39%). However, the version that emphasized the ghost-hunter's gadgets and jargon *did* influence beliefs. Compared to the participants who read the supernatural-themed story, the ones who read the scientific-sounding story expressed greater belief in haunted houses (50%) and greater faith in the scientific legitimacy of paranormal investigators (57%). In this study, reading about ghost-hunters wasn't enough, in and of itself, to bolster belief in the paranormal; instead, highlighting their terminology, technology, and "meticulous approach" was what did the trick.

Yet science can take away credibility as well as bestow it. Relative to the participants who read the scientific-sounding version of the story, those who read the third version—the one that debunked the ghost hunter's methods—were

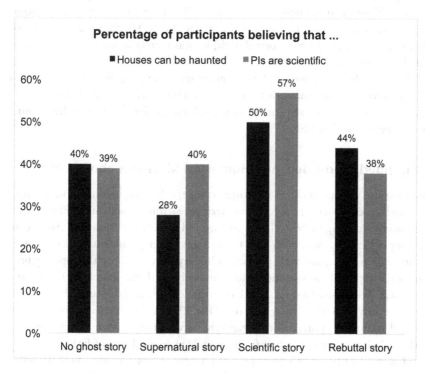

FIGURE 9.3 Beliefs about haunted houses and paranormal investigators, by news story condition (Paranormal news experiment, 2011)

less likely to believe in haunted houses (44%) and see paranormal investigators as scientific (38%). Adding a rebuttal from a scientist essentially neutralized the impact of cloaking paranormal investigations in the "trappings of science"—a result that highlights how skeptics can counter media claims about fringy phenomenon by challenging their basis in science.

When Paul published his study, it stirred up a bit of media discussion. Given that we're communication scholars, it was fun to see how news sources, skeptical forums, and other outlets framed the findings. One website, Giant Freakin Robot, ran a story under the headline, "Complex Words and Machinery Make Ghosthunters Seem Less Bullshitty" (a fair interpretation of the results).[42] Meanwhile, a conspiracy theory website quoted a Live Science story about the study and then argued that there was "something of the playfully absurd in a popular [news site] using scientific jargon seeking to influence public opinion by citing a professor on a study he conducted looking at popular media using scientific jargon to influence public opinion, and how citing a professor can mitigate that influence" (also a fair, if convoluted, point).[43] One clear pattern did emerge: the journalists who contacted Paul, from the *Wall Street Journal* reporter writing about haunted houses for sale to the local reporter writing about zombies, invariably asked whether he believed in ghosts. He always explained that he wasn't studying the paranormal itself; he was merely looking at the effects of media messages *about* the topic.

Some professors, however, have been eager to share their conclusions about the paranormal with journalists. In fact, our next study focused on the case of a psychologist who received national media attention for claiming to have found evidence of "psi," or ESP.

One-sided, Two-sided, and Humorous Messages about ESP

Among the various branches of fringe science, from cryptozoology to ufology, research on extra-sensory perception carries the strongest aura of credibility with the public. Judging by both Gallup polling and our own 2018 survey, Americans are more likely to believe in ESP than in any other paranormal phenomenon. This reflects a long and well-publicized history of research on the topic by professors at reputable universities. Most famously, perhaps, Joseph B. Rhine and Louisa E. Rhine used a special deck of cards (Zener cards) to study clairvoyance and telepathy at Duke University from the 1930s to the 1960s. Their research helped inspire the portrayal of parapsychology in *Ghostbusters*, but attempts by other scholars to replicate their findings yielded no evidence of ESP.[44]

The field of parapsychology received a new jolt of publicity in 2011 when Daryl Bem, an emeritus professor of psychology at Cornell University, published an article about ESP in the prestigious *Journal of Personality and Social Psychology*. The article presented what Bem claimed was experimental evidence for precognition (that is, the ability to sense the future), but many of his colleagues challenged his conclusions. One team of researchers published an article criticizing how Bem had interpreted his data, and other psychologists reported unsuccessful attempts to replicate his findings in new studies.[45]

Media outlets seized on the controversy: the *New York Times, Discover Magazine,* and *Wired* all reported on it, as did broadcast and cable television news programs. These outlets varied, however, in how they framed Bem's research. Some news organizations reported his findings without highlighting counterarguments from other researchers—as when MSNBC aired a January 23, 2011 interview of Bem himself. Other outlets balanced claims of evidence for ESP with scientific criticisms. For example, a January 6, 2011 CBS News story described Bem's results but then quoted a rebuttal from a group of psychologists.[46]

A handful of media outlets also took the opportunity to poke fun at Bem and his research. In some cases, the humor was subtle. Take the segment that *ABC News with Diane Sawyer* aired on January 6, 2011: after presenting an interview of Bem, it used a swirly graphic and a wacky sound effect in transitioning to a shot of a crystal ball. In other cases, the mockery was overt, as when Stephen Colbert devoted a January 27, 2011 segment of his comedy news program, *The Colbert Report*, to satirizing Bem's research. "I know what you're thinking," Colbert told viewers, "'Stephen, that's bullshit.' But on the other hand, I *know* you're thinking, 'Stephen, that's bullshit.'" The host did allow Bem to present his case as a guest on the program but introduced him with a joke about seeing the future: "Thank you for going to have spoken to me. You will have been a great guest."

Can such messages sway audience members' beliefs about ESP? To answer this question, Paul conducted an experiment in April 2012.[47] His undergraduate assistants recruited 446 fellow students to participate in the study, each of whom he randomly assigned to one of four groups. Those in the first group read a version of a real *New York Times* story (under the headline, "Journal to publish paper on ESP") edited so that it described Bem's research without presenting any criticism of it. Participants in the second group read a different version of the story ("ESP paper expected to outrage psychologists") that quoted a skeptical psychologist and pointed out that efforts to replicate Bem's findings had failed. Those in the third group read yet another version of the story ("You might already know this ...") that described Bem's history of performing "Kreskin-style magic acts for students and friends" and quoted the same skeptical psychologist as saying, "[Bem's] got a great sense of humor ... I wouldn't rule out that this is an elaborate joke." The participants in the final group read a story on an unrelated topic.

Of the participants in the no-ESP story group, 43% said they believed in ESP and 59% saw ESP researchers as "somewhat" or "very" scientific (Figure 9.4). These numbers reflect how widespread belief in ESP is among

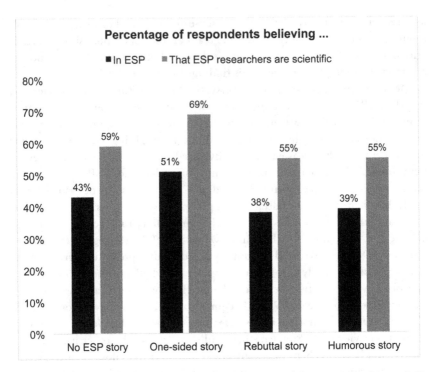

FIGURE 9.4 Beliefs about ESP and ESP researchers, by news story condition (ESP news experiment, 2012)

the public, including college students. Yet media messages that legitimize ESP research can push faith in its credibility even higher. Among the participants who read the one-sided story about Bem's findings, 51% believed in ESP and fully 69% saw ESP researchers as scientific.

By contrast, reading either of the other two versions led to *decreased* belief in ESP and *decreased* perceptions of ESP researchers as scientific. Of the participants who read the version that included a scientific rebuttal, only 38% believed in ESP and only 55% saw ESP researchers as scientific. The results were essentially identical for the humorous story: 39% of the participants who read it believed in ESP, and 55% of them saw ESP researchers as scientific. All this goes to show that Bem's critics had the opportunity to follow two different but equally effective paths in responding to his claims: by presenting scientific counterarguments or by treating them as a joke. More broadly, the findings from both the ghost-hunter experiment and the ESP experiment reveal that media messengers' choices about how to frame paranormal research help determine whether their stories encourage faith in or doubts about such research.

Foretelling the Future of Paranormal Media

Public belief in the paranormal shows no signs of fading anytime soon. Far from it: a series of national surveys sponsored by Chapman University from 2016 to 2018 found rising levels of belief in hauntings, ancient alien visitors, telekinesis, and Bigfoot.[48] Such views reflect a variety of social factors, but the media can play a crucial role in reinforcing them. Watching television—particularly paranormal-themed reality television—goes hand in hand with believing in fringy phenomena, and media messages that lend paranormal research an aura of legitimacy can bolster audience members' faith in aliens, ghosts, and ESP. By the same token, skeptical and satirical messages can undercut paranormal investigators' claims to scientific authority.

As for the future, it doesn't take psychic abilities to foresee that the media's fascination with the paranormal will continue. Indeed, media attention to UFOs spiked in 2019 with reports that US Navy personnel had seen what the military labeled "unidentified aerial phenomena." Newspapers such as the *Washington Post* and the *New York Times* ran articles about the sightings, television news shows replayed videos of them, and paranormal reality television producers rushed to create documentaries about them.

As in previous waves of paranormal hype, the resulting media messages presented a blend of credulous, skeptical, and satirical frames. True to form, the History Channel launched a series about the sightings, *Unidentified: Inside America's UFO Investigation*, that featured interviews with several high-profile UFO believers, including former US military intelligence official Luis Elizondo and musician Tom DeLonge. Moreover, some traditional news outlets joined in on the sensationalizing of the story. In an impressive bit of cross-promotion, a May 31, 2019 segment of ABC's *Good Morning America* even gave Elizondo and DeLonge the opportunity to promote the History Channel series along with their views on UFOs (both channels are owned by Disney).

Other media stories about the sightings took a more skeptical angle. For example, a CNN segment from June 2, 2019, featured an interview with Bill Nye, who argued that the pilots had probably seen secret military technology rather than alien visitors. Similarly, a September 19, 2019 NBC News story quoted astronomer Seth Shostak as saying that he saw "no reason to consider" the videos taken by the pilots to be "good evidence for 'alien visitation.'"

Late-night hosts, in turn, made wisecracks about the story. "The Navy was forced to address the [UFO] footage after it was released by a group called To the Stars Academy of Arts and Science, which was founded by former Blink-182 guitarist Tom DeLonge," Stephen Colbert explained to *Late Show* viewers on September 19, 2019. "Which might seem weird, but DeLong is just one of many 90s musicians doing trailblazing exploration of the paranormal. We were all indebted to the vigilance of the brave men and women over at the

Spin Doctors' Chupacabra task force." Colbert was joking, but ten days later the Travel Channel aired "Hunt for the Chupacabra," a new episode of its mystery-themed documentary series *Expedition Unknown*.

As DeLonge was spreading the word about UFOs to his millennial fanbase (and some Generation Xers as well), children's television shows were bringing paranormal themes to a new generation of viewers. Disney's *Gravity Falls* (2012–2016) led the way with its portrayal of two young investigators who chase cryptids and uncover conspiracies. Rival networks and streaming services soon followed with shows such as the Cartoon Network's *Infinity Train* (2019–2020) and Netflix's *The Hollow* (2018–2020). While these shows include echoes of *The X-Files* and *Lost*, they also fit into a long tradition of science-themed children's television—a genre that could shape young people's broader perceptions of scientists and scientific work.

Notes

1 Ackroyd, Dan, "About ghosts," *Huffington Post*, March 18, 2010, www.huffpost.com/entry/about-ghosts_b_327453; Galloway, Stephen, "The making of a comedy classic: Director Ivan Reitman spills the secrets behind the original *Ghostbusters*," *Hollywood Reporter*, July 15, 2016, www.hollywoodreporter.com/features/ghostbusters-director-ivan-reitman-making-911408; Iley, Chrissy, "Dan Ackroyd: A comedy legend's spiritual side," *Telegraph*, February 28, 2012, www.telegraph.co.uk/culture/film/9097816/Dan-Aykroyd-a-comedy-legends-spiritual-side.html.

2 VanHoose, Benjamin, "*Ghostbusters* Bill Murray, Dan Aykroyd and more remember Harold Ramis in virtual reunion," *People*, June 16, 2020.

3 Roach, Mary, *Spook: Science tackles the afterlife*, WW Norton & Company, 2006.

4 The Atlantic Paranormal Society, January 19, 2021, http://the-atlantic-paranormal-society.com.

5 Werts, Diane, "High spirits: Reality shows are alive with ghosts," *Newsday*, October 22, 2009, https://www.newsday.com/entertainment/tv/high-spirits-reality-shows-are-alive-with-ghosts-1.1539060.

6 Radford, Benjamin, "Ghost hunters' unscientific, win-win approach," Center for Inquiry, November 30, 2009, https://centerforinquiry.org/blog/ghost_hunters_unscientific_win-_win_approach.

7 Nickell, Joe, "Scientific investigation vs. ghost hunters," *Skeptical Inquirer*, September 28, 2011, https://skepticalinquirer.org/newsletter/scientific_investigation_vs-_ghost_hunters.

8 Moore, David W., "Three in four Americans believe in paranormal," Gallup, June 16, 2005, https://news.gallup.com/poll/16915/three-four-americans-believe-paranormal.aspx.

9 CBS News, "CBS News Poll: Workplace/marijuana/Hollywood celebrities," Social Science Research Solutions, Cornell University: Roper Center for Public Opinion Research, 2014; CBS News, CBS News Poll, Social Science Research Solutions, Cornell University: Roper Center for Public Opinion Research, 2017.

10 Saad, Lydia, "Americans skeptical of UFOs, but say government knows more," Gallup, September 6, 2019, https://news.gallup.com/poll/266441/americans-skeptical-ufos-say-government-knows.aspx.

11 Festinger, Leon, Henry W. Riecken, and Stanley Schachter, *When prophecy fails: A social and psychological study of a modern group that predicted the destruction of the world*, Harper Torchbooks, 1966.

12 Sparks, Glenn G., C. Leigh Nelson, and Rose G. Campbell, "The relationship between exposure to televised messages about paranormal phenomena and paranormal beliefs," *Journal of Broadcasting & Electronic Media* 41, no. 3 (1997): 345–359; Sparks, Glenn, and Will Miller, "Investigating the relationship between exposure to television programs that depict paranormal phenomena and beliefs in the paranormal," *Communication Monographs* 68, no. 1 (2001): 98–113.

13 Bader, Christopher D., Joseph O. Baker, and F. Carson Mencken, *Paranormal America: Ghost encounters, UFO sightings, Bigfoot hunts, and other curiosities in religion and culture*, NYU Press, 2017.

14 Dawkins, Richard, "Richard Dimbleby lecture," BBC TV, November 12, 1996; Yoo, Esther S., "Skeptics find fault with media portrayals of paranormal," *Harvard Crimson*, November 3, 1998, www.thecrimson.com/article/1998/11/3/skeptics-find-fault-with-media-portrayals.

15 Radford, Benjamin, "Are ghosts real? Evidence has not materialized," *LiveScience*, May 18, 2017, www.livescience.com/26697-are-ghosts-real.html; Smith, Alison, "TAPS vs. SAPS: The Atlantic Paranormal Society meets the Skeptical Analysis of the Paranormal Society," *Skeptic*, September 2006, www.skeptic.com/eskeptic/06-08-10.

16 Roach, *Spook*.

17 Rensberger, Boyce, "Magicians term Israeli 'psychic' a fraud," *New York Times*, December 13, 1975, www.nytimes.com/1975/12/13/archives/magicians-term-israeli-psychic-a-fraud.html.

18 Lagerfeld, Nathalie, "How an alien autopsy hoax captured the world's imagination for a decade," *TIME*, June 24, 2016, https://time.com/4376871/alien-autopsy-hoax-history.

19 Nisbet, Matt, "Cultural indicators of the paranormal," *Skeptical Inquirer* 26, no. 8 (2006), https://skepticalinquirer.org/exclusive/cultural_indicators_of_the_paranormal.

20 Nisbet, "Cultural indicators."

21 Nisbet, Matt, "Talking to heaven through television," *Skeptical Inquirer*, Mar. 13, 2001, https://skepticalinquirer.org/exclusive/talking_to_heaven_through_television/?/specialarticles/show/talking_to_heaven_through_television.

22 Nickell, Joe, "John Edward: Hustling the bereaved," *Skeptical Inquirer* 25, no. 6 (2001), https://skepticalinquirer.org/2001/11/john_edward_hustling_the_bereaved.

23 Nisbet, "Cultural indicators."

24 Radford, Benjamin, "John Edward's televised tragedy seance scrapped," *Skeptical Inquirer* 26, no. 1 (2006).

25 Sparks et al., "The relationship between exposure"; Sparks and Miller, "Investigating the relationship."

26 Sparks et al., "The relationship between exposure."

27 Sparks and Miller, "Investigating the relationship."

28 Nisbet, "Cultural indicators."

29 Gowen, Annie, "In Alexandria, exploring the city's ghostly side," *Washington Post*, May 4, 2004, https://www.washingtonpost.com/archive/local/2004/05/04/in-alexandria-exploring-the-citys-ghostly-side; Jordan, Pat, "The Dixie Ghostbusters," *New York Times*, December 11, 2011, www.nytimes.com/2011/12/11/magazine/the-dixie-ghostbusters.html

30 Black, Riley, "The idiocy, fabrications and lies of *Ancient Aliens*," *Smithsonian Magazine*, May 11, 2012, www.smithsonianmag.com/science-nature/the-idiocy-fabrications-and-lies-of-ancient-aliens-86294030; MonsterTalk, "Ancient alien astronauts: Interview with Ken Feder," *Skeptic*, July 27, 2011, www.skeptic.com/podcasts/monstertalk/11/07/27/transcript.

31 Ho, Rodney, "*Stranger Things* season 3 seen by 64 million in first month after release, most popular on Netflix," *Atlanta Journal-Constitution*, October 18, 2019, www.ajc.com/blog/

radiotvtalk/stranger-things-season-seen-million-first-month-after-release-most-popular-netflix/jGUCzwRoF36PxTHdqhLt2L.

32 Brewer, Paul R., "The trappings of science: Media messages, scientific authority, and beliefs about paranormal investigators," *Science Communication* 35, no. 3 (2013): 311–333.

33 *CSI: Crime Science Investigation*, "Ghosts of the Past" (Season 13, Episode 21); *The Simpsons*, "The Springfield Files," (Season 8, Episode 10).

34 See the Appendix for details.

35 Sparks, Glenn G., T. Hansen, and R. Shah, "Do televised depictions of paranormal events influence viewers' paranormal beliefs?" *Skeptical Inquirer* 18 (1994): 386–395.

36 Sparks, Glenn G., Cheri W. Sparks, and Kirsten Gray, "Media impact on fright reactions and belief in UFOs: The potential role of mental imagery," *Communication Research* 22, no. 1 (1995): 3–23.

37 Sparks, Glenn G., and Marianne Pellechia, "The effect of news stories about UFOs on readers' UFO beliefs: The role of confirming or disconfirming testimony from a scientist," *Communication Reports* 10, no. 2 (1997): 165–172.

38 Sparks, Glenn G., Marianne Pellechia, and Chris Irvine, "Does television news about UFOs affect viewers' UFO beliefs? An experimental investigation," *Communication Quarterly* 46, no. 3 (1998): 284–294.

39 Hornig, Susanna, "Television's NOVA and the construction of scientific truth," *Critical Studies in Media Communication* 7, no. 1 (1990): 11–23; Kirby, David A., *Lab coats in Hollywood: Science, scientists, and cinema*, MIT Press, 2011.

40 Barnett, Michael, Heather Wagner, Anne Gatling, Janice Anderson, Meredith Houle, and Alan Kafka, "The impact of science fiction film on student understanding of science," *Journal of Science Education and Technology* 15, no. 2 (2006): 179–191.

41 Brewer, "The trappings of science."

42 Venable, Nick, "Complex words and machinery make ghosthunters seem less bullshitty," *Giant Freakin Robot*, 2012, www.giantfreakinrobot.com/sci/complex-words-machinery-ghosthunters-bullshitty.html.

43 NWO Truth, October 29, 2012, https://nwotruth.org.

44 See, e.g., Cox, William S., "An experiment on extra-sensory perception," *Journal of Experimental Psychology* 19, no. 4 (1936): 429–437; Hines, Terence, *Pseudoscience and the paranormal*, Prometheus Books, 2010; Jastrow, Joseph, "ESP, house of cards," *The American Scholar* 8, no. 1 (1938): 13–22.

45 Ritchie, Stuart J., Richard Wiseman, and Christopher C. French, "Failing the future: Three unsuccessful attempts to replicate Bem's 'Retroactive Facilitation of Recall' Effect," *PloS one* 7, no. 3 (2012): e33423; Wagenmakers, Eric-Jan, Ruud Wetzels, Denny Borsboom, and Han van der Maas, "Why psychologists must change the way they analyze their data: The case of psi—comment on Bem (2011)," *Journal of Personality and Social Psychology* 100, no. 3 (2011): 426–432.

46 Katz, Neil, "Does controversial study prove ESP is real?" CBS News, January 6, 2011, www.cbsnews.com/news/does-controversial-study-prove-esp-is-real.

47 Brewer, Paul R., "How news about ESP research shapes audience beliefs," *Skeptical Inquirer*, September/October 2013: 41–43.

48 Chapman University Survey of American Fears, "Paranormal America 2018," Chapman University, https://blogs.chapman.edu/wilkinson/2018/10/16/paranormal-america-2018.

10

KID SCIENCE

Theme Song: Bill, Bill, Bill, Bill, Bill, Bill, Bill! Bill Nye the Science Guy!
Science rules!

—*Bill Nye the Science Guy* (1993–1998)

Dr. Heinz Doofenshmirtz: My name is Dr. Heinz Doofenshmirtz. My occupation:
evil scientist!
Judge: Mad scientist?
Doofenshmirtz: No. No. No, not mad scientist! I'm not angry. Evil scientist.
There's a difference.

—*Phineas and Ferb* (2012; Season 3, Episode 29: Norm Unleashed)

The first episode of the 1990s television show *Bill Nye the Science Guy* introduced its host with a barrage of graphics and a catchy theme song destined to become familiar to millions of American children. Over the course of half an hour, Nye explained the physics of flight by making balloon animals, paragliding over a lake, and blasting a hair dryer at his face. He also shared the screen with a "way cool scientist" (test pilot Suzanna Darcy), a team of young lab assistants, and a music video for a song titled "Smells Like Air Pressure" (spoofing "Smells Like Teen Spirit" by grunge rock band Nirvana). The episode ended with the host reviewing Bernoulli's principle and then shouting, "Gotta fly, Bill Nye the Science Guy!"

Nye had followed a winding path to national television. As a student at Cornell University, he took a course with famed astronomer Carl Sagan and

DOI: 10.4324/9781003190721-10

graduated with an engineering degree.[1] After working for Boeing, he quit to pursue a career as a comedian.[2] During the late 1980s, he created his "Science Guy" persona and teamed up with two local television producers to develop a concept for an educational children's show. They spent four years pitching the idea until they persuaded the National Science Foundation, PBS, and Disney to support it.

From 1993 to 1998, *Bill Nye the Science Guy* featured scientific demonstrations and profiles of real scientists alongside parodies of movies, television shows, and music videos, all edited together in a rapid-fire "MTV style." The "Science Guy" himself was an energetic host, alternately spouting corny jokes and enthusing about the wonders of science while wearing a bow tie and powder-blue lab coat. His formula worked: the show found a wide audience, and Nye became a pop culture icon. Despite the NSF's initial concerns that his quirky persona would reinforce stereotypes of scientists as "weird," surveys found that audience members were more likely to see scientists as "interesting, fun people" after watching the show."[3]

Since the end of his original series, Nye has expanded his television profile through interviews on news programs as well as appearances on shows such as *Dancing with the Stars* and *The Big Bang Theory*. Along the way, he has occasionally sparked criticism from both inside and outside the scientific community—for example, by debating evolution with creationist Ken Ham in 2014 and attending President Donald Trump's State of the Union address in 2018. Throughout all of this, he has remained one of the most prominent science communicators in the United States. When we asked a thousand Americans in October 2016 to name a scientist on television or in movies, Nye was easily the most popular choice: 17% of our respondents listed him, while no other person, real or fictional, topped 10%.[4]

Yet not all children's television offers depictions of science as glowing as the ones on *Bill Nye the Science Guy*. In contrast to educational programming, entertainment-oriented children's shows often portray science more ambivalently. For example, *Phineas and Ferb*, an animated Disney series that ran from 2007 to 2015, presents both flattering and unflattering images of scientists. On the one hand, the show's title characters are two ingenious and good-natured young stepbrothers who spend their summer vacation pursuing a dizzying variety of STEM-related projects, including (to quote the show's theme song) "building a robot," "finding a dodo bird," and "locating Frankenstein's brain." On the other hand, the show also features Heinz Doofenshmirtz, a self-proclaimed evil scientist whose schemes to "take over the tri-state area" are invariably foiled by Phineas and Ferb's pet platypus Perry.

Doofenshmirtz bears all the hallmarks of a classic stereotypical scientist: he has bulging eyes and wild hair, wears a lab coat, and speaks with a German-sounding accent. He's also a white man, just like Bill Nye. In fact, many of the scientists on children's television, real and fictional, share this demographic profile. Meanwhile, women scientists and scientists of color have historically occupied less visible roles on educational shows such as *Bill Nye the Science Guy* and, especially, entertainment-oriented children's shows such as *Phineas and Ferb*.

The messages about science in both genres may help mold young viewers' beliefs about *what* science is like and *who* can be a scientist. Such beliefs, in turn, can carry important consequences for science and society. A wide range of scientific fields will undoubtedly play key roles in shaping our future, but their appeal to new generations will depend in part on the public image of science itself. If young people see the profession in a favorable light and see themselves as potential scientists, they'll be more inclined to seek scientific education and careers.[5]

Perceptions of science may be particularly crucial when it comes to fostering a more diverse scientific community. As activists and leading scientific organizations have pointed out, recruiting a scientific community that mirrors our society will lead to new insights and help more of the potential scientists among us develop their talents.[6] Many sorts of obstacles contribute to the lingering inequities in science, but popular images of the profession could help to reinforce—or, alternatively, dismantle—these barriers.

The views of future *non*-scientists are also important to the ongoing relationship between science and broader society. If we hope to build support for scientific endeavors, bridge gaps between scientists and laypeople, and foster critical engagement with science, then we need to understand the childhood origins of public beliefs about science. The worldviews that young people hold often carry over into their adult lives and provide a foundation for their future judgments. Compared to adults, children also tend to be more open—or susceptible, depending on one's point of view—to social influence from many directions, including media messages.

In a series of studies, Jocelyn Steinke and Marilee Long have highlighted two theories that help explain this sort of influence when it comes to perceptions of science.[7] One is social cognitive theory, which emphasizes the effects of media models on young people's beliefs and aspirations.[8] The other is schema theory, which points to how media portrayals mold young people's stereotypes and self-images.[9] Both theories suggest that if children's shows present science as a dangerous, solitary endeavor for socially awkward oddballs, then young viewers will shy away from the topic. By the same logic, however, portrayals of scientists as relatable people doing interesting work could sway audience members toward engaging with science. For some viewers, such depictions may even foster a sense of "wishful identification" with scientist characters along with "possible selves" as potential scientists.[10] Encouraging these sorts of positive impressions has been a key goal of the educational television genre since the early years of television.

How Educational Children's Television Portrays Science

One of the first—and most successful—children's shows about science was *Watch Mr. Wizard*, which debuted on NBC in 1951. Host Don Herbert, a former air force pilot with a likable, down-to-earth persona, conducted simple but dramatic demonstrations with everyday objects to teach scientific principles to his young assistants, who often reacted by shouting, "Gee, Mr. Wizard!"[11] The series ran for 15 years, then returned for one season on NBC in 1971–1972 and underwent another revival as *Mr. Wizard's World* on the children's cable network Nickelodeon from 1983 to 1990. Herbert made a lasting impression on many of the people who grew up while his original show was on the air: in our October 2018 survey of the American public, 38% of the "silent generation" respondents (born pre-1945) and 31% of the baby boomers (born between 1945 and 1964) remembered watching *Mr. Wizard* as children.[12] By comparison, only 24% of Generation Xers (born between 1965 and 1980) and 8% of millennials (born after 1980) had seen the show.

Most members of Generation X (including us) are more familiar with *Schoolhouse Rock!*, a series of short musical cartoons that premiered on ABC in 1973. After three seasons' worth of songs about multiplication, grammar, and American history, the series turned to STEM topics in 1978 and 1979. The cartoons in the Science Rock season included "Energy Blues" (about conservation), "Victim of Gravity" (performed by the Tokens, a doo-wop vocal group), and "Electricity, Electricity" (the catchiest song of the set, in our opinion). Among the Gen Xers in our 2018 survey, fully 62% remembered watching *Schoolhouse Rock*. Almost half of the millennials (48%) and more than a third of the boomers (36%) did so, as well. Our October 2016 survey also found that 10% of all respondents—and 18% of Gen Xers—specifically reported watching "Interplanet Janet," a *Schoolhouse Rock!* cartoon that tells the story of "a galaxy girl, a solar system Ms. from a future world; she travels like a rocket with her comet team, and there's never been a planet Janet hasn't seen."

PBS launched two science-themed children's programs during the 1980s: *3-2-1 Contact*, which ran from 1980 to 1988, and *Newton's Apple*, which aired from 1983 to 1998. Both shows emphasized how scientific principles work in real-world settings, from recording studios to roller coasters.[13] They also tried to break the demographic mold of science set by *Watch Mr. Wizard* and other early educational shows featuring white men. "Too many children think that scientists are all middle-aged white males in laboratory coats," *3-2-1 Contact* director Edward Atkins explained in a 1983 interview. "We want to introduce them to other kinds of scientists—women, minorities, people using science in daily life."[14] For us, the most memorable part of *3-2-1 Contact* (apart from the funky theme song) was "The Bloodhound Gang," a series of segments about a diverse team of boy and girl detectives who use science to solve mysteries.[15] Of the respondents in our 2018 survey, 10% remembered watching *3-2-1 Contact*, and the same percentage recalled watching *Newton's Apple*. Not surprisingly, Gen Xers were more likely than everyone else to have seen each show.

The 1990s brought a new wave of science-themed educational shows, driven in part by the federal government's decision to tighten its enforcement of regulations on children's television.[16] One new entry in the genre was *Beakman's World*, which—like *Bill Nye the Science Guy*—featured a zany host with a brightly colored lab coat (a neon green one, in this case) and a visual style influenced by MTV. In the character of Beakman, actor/performance artist Paul Zaloom demonstrated scientific principles through experiments and stunts, aided by an anthropomorphic lab rat and a revolving-door cast of female assistants. The show ran from 1992 to 1998 on The Learning Channel and in syndication, to acclaim from television critics and the National Science Foundation.[17]

Bill Nye the Science Guy was even more successful, establishing its host as the Mr. Wizard for a new generation.[18] Almost half (48%) of the respondents in our October 2018 survey said they had watched Nye's series. The show also stood out for its cross-generational appeal: 28% of silent generation respondents, 31% of boomers, 40% of Gen Xers, and 76% of millennials recalled seeing the program. Furthermore, another one of our surveys—this one conducted in July 2016—found that Nye himself was both well-known and widely liked. Three-fifths (60%) of the respondents held a favorable opinion of him whereas only one-fifth (19%) viewed him unfavorably (the remaining 21% had no opinion).[19]

The Magic School Bus took a different approach to science education: instead of featuring a male host like Beakman or Bill Nye in a live-action laboratory, it presented the animated adventures of science teacher Valerie Felicity Frizzle and her students. For example, one episode follows Ms. Frizzle as she shrinks her bus to microscopic size and drives her class into the body of Ralphie, a sick student. The show's hero finds everything fascinating, even when white blood cells attack the bus (everyone makes it away safely). *The Magic School Bus* originally aired on PBS from 1994 to 1997 and found a wide audience among millennials: four-fifths (80%) of the ones in our October 2018 survey had watched it.

Just as the mini-boom in children's science television peaked in 1994, Marilee Long and Jocelyn Steinke conducted a content analysis of four live action shows in the genre: *Newton's Apple*, *Beakman's World*, *Bill Nye the Science Guy*, and *Mr. Wizard's World*.[20] The researchers found that these shows never reinforced stereotypes of scientists as evil or violent, and only rarely presented science as mysterious or magical. Instead, all four programs emphasized how science is both part of everyday life and (particularly in the case of *Bill Nye* and *Beakman's World*) fun. The shows sometimes portrayed scientists as eccentric (again, most notably in the cases of Nye and Beakman) and science as dangerous (both Beakman and Nye were fond of explosions) but also presented science as being "for everyone" by featuring guest scientists and lab assistants who were diverse in terms of age, gender, and race (though Mr. Wizard, Nye, and Beakman themselves were all white men in their 40s or older).

Twelve years later, Long led another study looking at portrayals of scientists in children's educational television.[21] One of the shows in this follow-up analysis was *Bill Nye the Science Guy*, which had ended production but was still airing in reruns. Another was *DragonflyTV*, a PBS series about "real-life science investigations featuring everyday kids" that aired from 2002 to 2008.[22] Yet another was *MythBusters*, a demonstration-focused program starring two snarky special effects experts that ran on The Discovery Channel from 2003 to 2016. Long and her team found that these programs seldom portrayed scientists as working alone and virtually never stereotyped them as being nerdy or violent. Far more often, the three shows presented likable, intelligent people working together to study science.

Educational programs featuring similar depictions have flourished over the past decade, with PBS remaining the clear leader in the genre. Its science-themed animated shows include *Sid the Science Kid*, which follows the young title character and his classmates as they learn about scientific principles, and *Dinosaur Train*, which depicts dinosaurs exploring topics in paleontology. PBS also distributes *SciGirls*, a series targeting tween girls that blends live action and animation, and *Wild Kratts*, a series featuring a pair of real-life zoologists and their cartoon alter egos. In 2019, one of the commercial broadcast networks, CBS, began airing *Mission Unstoppable*, a program that profiles women in STEM. Produced by Geena Davis (an actor and founder of the Geena Davis Institute on Gender and Media) and hosted by Miranda Cosgrove (of *iCarly* and *School of Rock* fame), it has featured guests such as Jacqueline "the STEM Queen" Means, an 18-year old Black neuroscience student who works to promote STEM engagement among children in her Wilmington, Delaware community.[23] For its part, the Disney Junior cable channel has run programs such as *The Octonauts*, a British-produced animated series about anthropomorphic animals who study marine biology.

Beyond traditional television, the rise of streaming video services has created new platforms for science-themed programming. For example, Netflix revived two icons from 1990s television in 2017: Bill Nye in *Bill Nye Saves the World* and the Frizzle family in *The Magic School Bus Rides Again*. The following year, the platform launched *Brainchild*, a demonstration-based show that follows the template of *Watch Mr. Wizard*, *Beakman's World*, and *Bill Nye the Science Guy* but is hosted by a woman of color in her twenties, Sahana Srinivasan.[24] In 2020, Netflix launched *Emily's Wonder Lab*, another science show for children featuring demonstrations and experiments. Its host, Emily Calandrelli, was visibly pregnant throughout the filming of the program's debut season. "I'm especially proud to be a pregnant woman doing science on a platform like Netflix," she tweeted on August 11, 2020. "Personally, I think this sends a very cool message as to WHO is welcome in STEM. These careers aren't typically very welcoming to families, particularly women who want families."

Educational Science TV and Perceptions of Scientists

Building on their research into how educational programming portrays science, Jocelyn Steinke and her collaborators have developed a two-part framework for understanding its impact on viewers. The first plank is social cognitive theory, famously tested by psychologist Albert Bandura through an experiment where he showed children videos of another child playing with a doll.[25] When he gave the participants toys, the ones who had watched the media model attack the doll were more likely to engage in violent play than the ones who watched the model play with the doll quietly.

As Steinke points out, the logic of social cognitive theory should extend to media portrayals of science: when young people come into "vicarious contact" with a television scientist, the nature of the portrayal may influence their attitudes toward real-world science.[26] Positive role models of scientists should encourage viewers to engage with the subject, whereas negative ones should do the opposite. Furthermore, young people who form a sense of wishful identification with a media model—such as a television scientist—should be especially likely to pattern their own aspirations and choices after that model.[27] In other words, watching educational science programming can make viewers want to be like Bill Nye or Ms. Frizzle.

The second piece of the framework developed by Steinke and her colleagues is schema theory. Sandra Bem and other psychologists have found that children typically develop mental structures, or schemata, about gender—such as "pink is for girls, blue is for boys" or "men are tougher than women"—at an early age and then use them to interpret the world.[28] Such beliefs provide the foundations for gender stereotypes, including stereotypes related to science.

Media messages, in turn, play an important role in shaping young people's schemata—particularly of groups, such as scientists, that children seldom encounter in their own social life. In the absence of real-world examples, television portrayals provide a ready source of stereotypes about scientists.[29] The same portrayals also shape young viewers' own possible selves, their visions of who they can or cannot become.[30] After all, wanting to be like Bill Nye or Ms. Frizzle isn't necessarily enough to follow in their footsteps; viewers may need to see *themselves* in media models, too.

Guided by social cognitive theory and schema theory, we used the results from our October 2018 survey of the US public to explore how watching educational children's television is linked to perceptions of science. To capture respondents' viewing history, we combined their answers to two questions. The first asked, "When you were growing up, how often did you watch educational TV shows about science?" Of the respondents, 26% said "often," 43% said "sometimes," 23% said "rarely," and 7% said "never." The second question was whether respondents had watched *Bill Nye the Science Guy*.

One pattern that stood out in the results was the link between watching children's educational television programs and seeing scientists as "dedicated people who work for the good of humanity." Around half (47%) of the most devoted viewers—the ones who said they had often watched educational science shows in general *and* had watched Bill Nye in particular—strongly agreed with this description of scientists (Figure 10.1). The percentage was 13 points lower among respondents who had only occasionally watched the same programs and 18 points lower among those who had seldom or never done so. In short, watching positive role models of scientists on children's television shows went hand in hand with seeing the entire profession in a positive light.

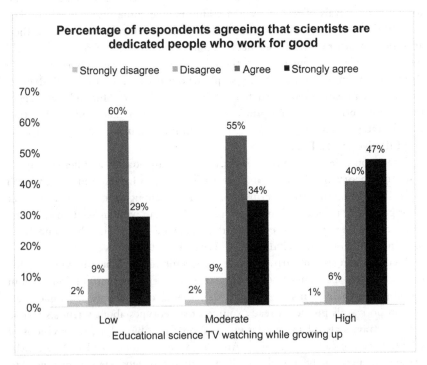

FIGURE 10.1 Perceptions of scientists as good, by educational TV watching (Cooperative Congressional Election Survey, 2018)

Vicarious contact with the likes of Bill Nye, Mr. Wizard, Beakman, and Ms. Frizzle may also help dispel negative stereotypes of scientists. As a case in point, faithful viewers of educational science television were particularly likely to *reject* the notion that "scientists tend to be odd and peculiar people." Almost a quarter (24%) of them strongly disagreed with this statement; among everyone else, the percentage was around ten points lower (Figure 10.2). Similarly, devoted viewers were especially likely to reject the stereotype of the solitary scientist. A third of them (33%) strongly disagreed that "scientists tend to work alone," double the percentage among everyone else (Figure 10.3).

All these patterns held even when we statistically accounted for other factors, including demographics, overall television viewing, and science fiction television viewing. Interestingly, however, we didn't find any link between children's educational science television viewing and perceptions of how dangerous

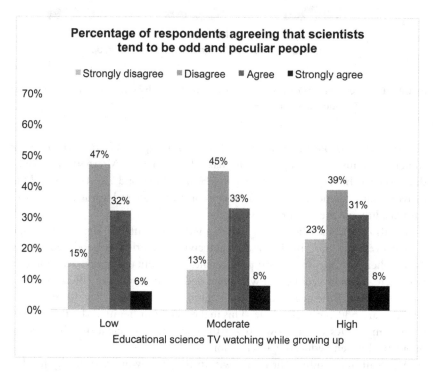

FIGURE 10.2 Perceptions of scientists as odd and peculiar, by educational TV watching (Cooperative Congressional Election Survey, 2018)

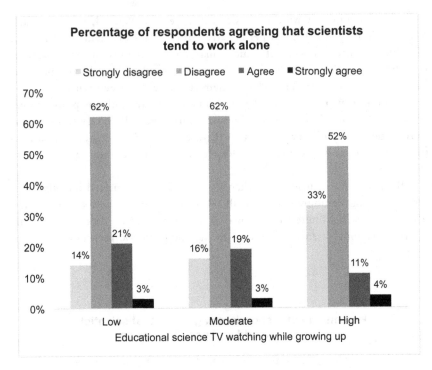

Percentage of respondents agreeing that scientists tend to work alone

FIGURE 10.3 Perceptions of scientists as working alone, by educational TV watching (Cooperative Congressional Election Survey, 2018)

scientific work is. Perhaps this type of programming simply doesn't influence audience members' beliefs about the hazards of science. Alternatively, maybe shows such as *Bill Nye the Science Guy*, *Beakman's World*, and *The Magic School Bus* deliver a more nuanced message: science is probably safe so long as you're with a trusted adult and wear your safety goggles.

We also explored whether the patterns we found differed across age groups. The answer was "no" in the case of perceiving scientists as good people, but "yes" when it came to stereotyping scientists as solitary geeks. Respondents who were relatively young *and* who were dedicated viewers of educational science television while growing up stood out for how strongly they disagreed that scientists are odd and work alone. This makes sense: millennials should hold the freshest memories of the down-to-earth, collaborative scientists who tend to populate children's educational television.

All in all, our findings fit neatly within the framework laid out by Jocelyn Steinke and her colleagues. Watching educational science television during childhood dovetailed with seeing scientists as working for good, as *not* working alone, and as *not* being odd and peculiar—patterns that reflect how shows such as *Bill Nye the Science Guy* present positive role models of scientists and challenge some common stereotypes of the profession. Our results here are silent as to whether these programs can also counter demographic stereotypes of scientists, but we'll revisit that issue once we've explored another key television genre: entertainment children's television programming, which is driven less by the goal of promoting science engagement and more by the goal of generating revenue through high ratings.

How Entertainment Children's Television Portrays Science

Unlike their counterparts in educational television, the scientists in children's entertainment television have historically been a mixed bag of mentors, heroes, and villains. When the broadcast networks began airing blocks of Saturday morning cartoons in the 1960s and 1970s, they ran shows featuring all three of these archetypes.[31] The mentors included Mr. Peabody from *The Bullwinkle Show* (1961–1964), a talking dog who leads a boy named Sherman on adventures through time with the WABAC machine, and Dr. Benton Quest from *Jonny Quest* (1964–1965), the scientist father of the show's young title character. Mr. Spock from the animated version of *Star Trek* (1973–1974) was typical of the scientist as hero: as first officer of the starship *Enterprise*, he uses his knowledge to fight evil and help people in need. Meanwhile, *Super Friends* (which originally aired in 1973–1974 and underwent numerous revivals) pitted superheroes such as Batman, Wonder Woman, and Superman against a seemingly endless supply of sinister or misguided scientists with names such as Professor Baffles and Dr. Thinkquick.

The Saturday morning cartoons of the 1980s and early 1990s featured new evil scientists along with new heroic ones. Among the villains were Dr. Claw, nemesis of the hero from *Inspector Gadget* (1983–1986), and Mister Sinister from *X-Men: The Animated Series* (1992–1997). The heroic scientists included the paranormal investigators of *The Real Ghostbusters* (1986–1991) and Donatello, the purple-masked member of the *Teenage Mutant Ninja Turtles* (1987–1996). All four shows drew high ratings, but changes in the media industry over the course of the 1990s led the broadcast networks to drop their Saturday morning cartoon blocks.

Fortunately for animation fans, cable networks such as the Cartoon Network, Nickelodeon, and the Disney Channel filled this vacuum with their own shows—which, like their broadcast predecessors, depicted scientists in a mix of roles. Sometimes these scientists were sympathetic, if flawed, young protagonists. For example, *Dexter's Laboratory* (1996–2003) portrays its title character, who always wears glasses and a white lab coat, as a socially awkward but good-hearted genius. Similarly, *The Adventures of Jimmy Neutron* (2002–2006) follows the life of a youthful inventor who gets himself into—and out of trouble—with his amazing devices.

Other cartoon scientists from the late 1990s and early 2000s fit the mentor role. Consider Professor Utonium, the lab-coated creator/father of young superheroes Blossom, Buttercup, and Bubbles from *The Powerpuff Girls* (1998–2005). He may be a bit clueless, particularly around women, but he's a kind and devoted parent. Dr. Nora Wakeman plays a similar role on *My Life as a Teenage Robot* (2003–2005) as the inventor/mother-figure to XJ-9, a girl-like automaton who renames herself Jenny Wakeman. The elder Wakeman checks many of the boxes on the stereotypical scientist list with her spiky white hair, glasses, vaguely European accent, and eccentric personality, though she breaks the historical stereotype of the male scientist.[32] *Danny Phantom* (2004–2007) also features scientists as parents to the main character: in this case, they're paranormal investigators who don't realize that their own son has ghostly powers.

As for the evil scientist archetype, it provided the model for many villains on turn-of-the-millennium toons. For example, a brilliant mutated monkey named Mojo Jojo is the arch-nemesis of *The Powerpuff Girls*, and paranormal researcher Vlad Masters is the main antagonist of *Danny Phantom*. The secret agent-themed *Kim Possible* (2002–2007) provides a particularly memorable example of the trope in Dr. Drakken, a spiky-haired, blue-skinned mad scientist who constantly schemes to take over the world. His plans tend to fail spectacularly due to their obvious flaws and needless complexity—as his sarcastic assistant Shego is quick to remind him.

Marliee Long and her colleagues analyzed four cable network cartoons as part of their 2006 content analysis: *Dexter's Lab*, *Kim Possible*, *The Adventures of Jimmy Neutron*, and *Danny Phantom*.[33] These shows occasionally presented scientists as violent (Dr. Drakken likes ray guns and doomsday devices, while even Dexter occasionally wields his wrench in combat) and as working alone (Dexter's preferred mode of research, though his sister often disrupts his efforts) but seldom depicted them as nerdy or geeky.

The mixed portrayals of scientists on the Cartoon Network, Nickelodeon, and the Disney Channel have continued since 2007. Indeed, that was the year *Phineas and Ferb* introduced its likable young title characters along with the hapless Heinz Doofenshmirtz (who even has his own villainous jingle: "Doofenshmirtz evil incorporated!"). Other children's animated shows from the past decade have featured a variety of colorful scientists—literally so in the case of *Adventure Time*'s (2010–2018) Princess Bubblegum, the bright pink ruler of the Candy Kingdom. She's a benevolent monarch, though she occasionally causes chaos with her experiments. Varrick, an inventor from *The Legend of Korra* (2012–2014), is more ethically ambiguous: accompanied by his long-suffering assistant Zhu Li, he alternates between helping and hindering the show's heroes while plotting to make money. *Gravity Falls* (2012–2016) portrays yet another complex cartoon scientist in the form of Stanford Pines, who carries on a destructive feud with his twin brother Stanley but also acts as a mentor to young paranormal researcher Dipper.

In 2016, we surveyed 317 college students to find out what they thought about some of the most iconic scientists from animated cable television. A huge majority (87%) of our respondents held a favorable view of Jimmy Neutron from *The Adventures of Jimmy Neutron*, and Dexter from *Dexter's Laboratory* was also popular (66% held a positive opinion of him). Respondents were less familiar with Professor Utonium from *The Powerpuff Girls*, but the ones who knew him saw him favorably by a three-to-one margin (45% to 15%).

Meanwhile, a pair of ostensible villains inspired ambivalent reactions. The respondents who were familiar with *Kim Possible*'s Dr. Drakken had split opinions of him (33% favorable versus 37% unfavorable), and the same was true for *Phineas and Ferb*'s Dr. Doofenshmirtz (38% versus 33%). These divided views presumably reflect the characters' statuses as comic antagonists who never actually hurt anyone. Doofenshmirtz is an especially pathetic figure: he's suffered a lifetime of indignities, beginning when his father named the family dog, "Only Son."

Testing the Effects of Cartoon Science on Young Viewers

Given how entertainment children's television programs present relatively mixed messages about science, we shouldn't expect these shows to affect perceptions of scientists in simple, one-sided ways. Young people who tune in to the Cartoon Network, Nickelodeon, or the Disney Channel may see positive role models like Professor Utonium and relatable peers like Jimmy Neutron, but they'll also come across villains like Dr. Drakken. Similarly, these viewers will encounter characters whose portrayals reinforce stereotypes of scientists as odd or awkward, like Dr. Doofenshmirtz, along with characters whose depictions challenge such stereotypes, like Phineas and Ferb.

In keeping with such mixed messages, our surveys showed no clear links between past viewing of children's entertainment television and perceptions of scientists. For example, the respondents in our 2016 student survey who recognized Dexter, Professor Utonium, Jimmy Neutron, Dr. Drakken, and Dr. Doofenshmirtz were no more—or less—likely than everyone else to see scientist as being good, or as being odd, or as working alone. Likewise, the respondents in our national 2018 survey who had watched *Dexter's Laboratory* (30% of the sample) and *The Powerpuff Girls* (27%) didn't differ from non-viewers in what they thought about scientists.

Even if watching entertainment children's television doesn't yield a consistent *overall* effect on these perceptions, however, watching *specific* portrayals of cartoon scientists could still influence how audience members perceive scientists. For example, young viewers could be inspired by watching *Adventure Time*'s Princess Bubblegum save the day with her inventions. Alternatively, they could be put off from science by seeing Dexter isolate himself in his lab or by watching Dr. Drakken embarrass himself through his incompetent scheming.

To test these sorts of possibilities, we conducted an experiment in 2019 where participants watched different clips from *Phineas and Ferb*. Our sample included 303 college students, most of whom were still teenagers at the time.[34] These participants didn't fall into *Phineas and Ferb*'s primary demographic of school-age children, but they fit squarely within the secondary audience targeted by the show's pop-culture references and satirical humor. Almost all the students (95%) had seen the series, and around three-fourths (72%) had seen it many times.

We randomly assigned our participants to four groups. One of these, the control group, didn't watch a clip from *Phineas and Ferb*. Instead, they watched a scene from another cartoon, *SpongeBob SquarePants*, featuring a pair of non-scientist characters (Squidward and SpongeBob himself).

A second group, the "evil scientists" condition, watched an edited version of a *Phineas and Ferb* episode ("A Hard Day's Knight") in which secret agent Perry the Platypus infiltrates a convention of evil scientists to foil Dr. Doofenshmirt'z latest scheme. This storyline portrays Doofenshmirtz and his fellow convention-goers as eccentric and villainous, though it doesn't depict them as loners; in fact, Doofenshmirtz is thrilled to meet his own role model and says it makes him "happy to be surrounded by such *evil*."

Another group, the "mixed scientists" condition, watched an edited version of a *Phineas and Ferb* episode ("Agent Doof") about Doofenshmirtz working with Perry to defeat an evil scientist named Dr. Diminutive. In this storyline, Doofenshmirtz attempts to change his ways—he even signs an "I give up evil" affidavit—but proves to be terrible at working with others. He's ultimately fired from Perry's spy agency for violating a long list of human resources policies.

The final group, the "good scientists" condition, watched a clip focusing on Phineas and Ferb. In this scene (from the episode "Chronicles of Meap"), the two young scientists come to the aid of an adorable alien who crash-lands in their backyard while they're testing a pair of electronic baseball mitts. The stepbrothers use two inventions to help the stranded extraterrestrial: a cybernetically-controlled mechanic's platform and a GPS "cute-tracker."

Afterward, we asked all four groups a series of questions about scientists. The answers revealed that watching the *Phineas and Ferb* clips didn't matter for some perceptions. For example, participants in every group tended to agree that scientists are dedicated people who work for the good of humanity (Figure 10.4).

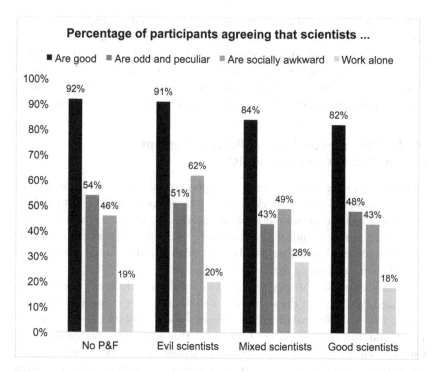

FIGURE 10.4 Perceptions of scientists, by *Phineas and Ferb* watching condition (*Phineas and Ferb* Experiment, 2019)

Nor did the clips have any clear impact on perceptions of scientists as odd and peculiar.

At the same time, the *Phineas and Ferb* videos did sway other beliefs about scientists. Most strikingly, the participants who watched the "evil scientists" clip were almost 20 percentage points more likely than those who watched the "good scientists" clip to see scientists as being socially awkward (62% versus 43%). This makes sense given the portrayals in the two clips: the first one shows Doofenshmirtz rambling obliviously to a poorly disguised Perry, whereas the second shows Phineas and Ferb engaging in normal, friendly conversations.

The "mixed scientist" video also had one clear effect: its portrayals of Dr. Diminutive's antisocial scheming and Dr. Doofenshmirtz's poor teamwork evidently encouraged viewers to endorse the image of the solitary scientist. The participants who watched this clip were more likely than the control participants to say that scientists tend to work alone (28% versus 19%).

In sum, the results of our experiment offer a mixed picture of *Phineas and Ferb*'s effects. Sometimes videos from the show reinforced stereotypes of scientists; sometimes they didn't. We suspect the same would be true for other children's entertainment cartoons, from *Dexter's Laboratory* to *Gravity Falls*— though we also suspect that the impact of these shows might be stronger among younger, relatively impressionable children than among college students.

Media Messages and Young People's Perceptions of Who Can Become a Scientist

So far, we've focused on how educational programming and entertainment-oriented cartoons can influence beliefs about what scientists are like. Yet children's television could also shape young viewers' perceptions of who scientists are—and who can become one. Indeed, a widespread concern in the scientific community is that stereotypical media portrayals of scientists will discourage girls and children of color from seeing themselves as scientists.

With good reason: throughout most of its history, children's television has reinforced the image of the scientist as a white man. For example, Steinke and Long's 1994 content analysis of *Mr. Wizard's World*, *Newton's Apple*, *Beakman's World*, and *Bill Nye the Science Guy* found that male scientists outnumbered female scientists on these educational programs by a two-to-one margin.[35] Furthermore, the shows often portrayed women and girls as students, apprentices, lab assistants, or science reporters, rather than full-fledged experts.

When Long and her colleagues looked at *Newton's Apple*, *Beakman's World*, *Bill Nye the Science Guy*, and *The Magic School Bus* two years later, they found a closer gender balance.[36] This time, roughly half the scientists were men or boys and half were women or girls. The shows also depicted male and female scientists as equal in status and gave them equal screen time. Less encouragingly, the white scientists on children's educational television averaged almost twice as much screen time as the scientists of color.

In 2006, Long and her colleagues analyzed the demographics of science on both educational and entertainment-orientated programs.[37] The results revealed a dramatic split between the genres. The two NSF-funded educational shows, *Bill Nye the Science Guy* and *DragonflyTV*, achieved gender parity in their portrayals of science (though the most prominent scientist, Nye, was a man). Meanwhile, entertainment programs such *Dexter's Laboratory*, *The Adventures of Jimmy Neutron*, and *Kim Possible* featured twice as many male scientists as female ones (cases in point: Dexter, Jimmy Neutron, and Dr. Drakken). Long's team also found that nearly three-fourths of the scientists on children's television were white (including Nye, Dexter, and Neutron).

Phineas and Ferb, which premiered a year later, follows the same patterns. Three of the central male characters—Phineas, Ferb, and Doofenshmirtz—are scientifically inclined, whereas the lone female lead, Candace, is more interested in dating boys, buying clothes, talking on her phone, and "busting" her younger brothers. "When it comes to mad science," comics critic Margaret O'Connell writes, "the girls and the women in the cast tend to be relegated to the roles of admiring and reasonably scientifically literate helpmeet … or memorable but rarely seen guest star."[38] The main scientist characters in *Phineas and Ferb* are also uniformly white. The supporting cast does include Baljeet, a South Asian math whiz, but some observers have argued that his depiction reflects stereotypes of Indian Americans as nerdy.[39]

Nor is *Phineas and Ferb* an exception in pigeonholing scientists by gender and race. When the Geena Davis Institute analyzed children's television shows from 2007 to 2017, it found that 59% of the STEM characters were male and 72% were white.[40] Moreover, women scientists on these programs tended to work in the life sciences rather than other STEM fields, a pattern that mirrors—and could reinforce—historical disparities in the profession.[41]

In conjunction with their study of children's television portrayals, the Geena Davis Institute conducted a nationally representative survey of middle school girls, high school girls, and women in college. An overwhelming majority of the respondents said that seeing girls and women as STEM characters on television was an important influence on their own choice of whether to pursue a STEM career. Almost three-fourths cited McKeyla McAlister, a character on the live action tween show *Project Mc²*, as an inspiration for pursuing STEM, and slightly more than half said the same of the main character from the animated children's series *Doc McStuffins*.

Other research reveals how media portrayals can shape young viewers' gender stereotypes of scientists. In one study, Steinke and her colleagues asked hundreds of seventh graders to draw a scientist.[42] Half the girls drew a woman, compared to only 13% of the boys. When asked where they got their ideas for these drawings, the students cited television shows and movies as their top source. Not surprisingly, the students who relied on media portrayals for inspiration were especially likely to draw stereotypical male scientists with wild hair, glasses, and lab coats.

A few years later, Steinke and her team explored how children's television influences young viewers' possible selves—that is, their images of who they hope, or fear, they can grow up to be.[43] The researchers showed a group of seventh graders clips from a variety of television programs, including educational shows (*Bill Nye the Science Guy*, *DragonflyTV*, and *MythBusters*) and entertainment shows (*Dexter's Lab*, *The Adventures of Jimmy Neutron*, *Kim Possible*, and *Danny Phantom*). Before watching these clips, the boys in the study were more likely than the girls to see themselves as being good at science *at the present*. However, viewing the clips encouraged both boys and girls to see themselves as being good at science *in the future*.

A follow-up analysis by Steinke and her colleagues looked at whether young viewers wanted to be like the scientists in the same television clips.[44] Across the board, the boys in the study were more likely to identify with male scientist characters (including Dexter, Jimmy Neutron, and Dr. Drakken) than with female scientist characters. Meanwhile, the girls expressed greater "wishful identification" with female scientists than with male scientists portrayed as working alone (such as Dexter) or trying to dominate other people (such as Drakken). Such patterns may reinforce gender disparities in who wants to become a scientist, particularly given how male scientists outnumber female scientists on children's television.

A recent experiment conducted by Bradley Bond shows that stereotypes in children's television can even influence elementary school students.[45] In this study, a group of girls between the ages of six and nine watched television clips of girls engaging in gender-stereotypical activities: meeting a fairy godmother, talking about how to impress boys, and discussing what to wear to a dance. Viewing the clips led the students to express greater interest in gender-stereotypical careers, such as teacher, florist, or stay-at-home mom. The girls who saw the stereotypical videos were also more likely to draw male scientists, compared to girls who didn't watch any clips. By contrast, watching counter-stereotypical clips of girls doing STEM activities—such as building robots or rockets—had no effect on viewers' career interests or drawings of scientists.

Putting these pieces together reveals a troubling picture of how gender stereotypes in children's television programs can sway perceptions of science. Such shows often reinforce the message that scientists are men, and young viewers pick up on this. Nor is that the end of the story. Similar dynamics could play out when it comes to the effects of other stereotypes in science-themed children's television programs. Given that these programs tend to depict scientists as white, young viewers of color may come away with the impression that real-world science will not welcome them. Girls of color, in particular, may experience a vicarious version of the "double bind" that results from intersecting gender- and race-based discrimination within science. On top of that, children's television typically portrays scientists as heterosexual and able-bodied. Hardly any cartoon scientists have visible disabilities, and the love interests for characters such as Professor Utonium, Phineas, and Dr. Doofenshmirtz are almost invariably members of the opposite sex.[46] All of these patterns may help replicate existing disparities in who pursues a life of science.

Still, children's television—particularly educational programming—has shifted toward greater balance, at least when it comes to portraying women scientists. Moreover, many young viewers do notice the exceptions among the mostly white, male world of cartoon science, such as Doc McStuffins (who is a Black girl). Which raises a broader question: what's the best way, if any, to counter stereotypes of scientists in children's media?

Challenging—and Reimagining—Portrayals of Science in Children's Media

One potential approach for dispelling such stereotypes is to educate young people to think critically about what they see and hear in the media. "Teachers exert considerable influence on young adolescents," write Lisa Ryan and Jocelyn Steinke. "It is critical during these adolescent years for science teachers to address and challenge students' stereotypes of science and scientists in order to promote the possibility of science careers as options for all students."[47]

As education scholar Thomas McDuffie points out, however, teachers may need to reflect on their own stereotypes of scientists before they can lead students in dissecting media messages.[48] When he asked a group of teachers and teachers-in-training to draw scientists, many of the respondents gave him pictures of lab-coated, bespectacled, wild-haired white men. To counter these stereotypes, McDuffie advocates not only holding class discussions but also inviting scientists as guest speakers and taking field trips to scientific workplaces to provide young people with real-life role models of scientists.

In some cases, education-based approaches can work. Our favorite success story along these lines is the field trip program at Fermilab, a particle physics laboratory.[49] Visiting the facility and meeting its researchers inspired a number of the seventh-grade girls in the program to switch from drawing stereotypical male scientists (complete with glasses, wild hair, and lab coats) to drawing less-stereotypical women.[50] Still, the same case illustrates the limits of educational programs. The Fermilab trip didn't influence how any of the seventh-grade boys in the program envisioned science: they all draw male scientists both before and after their visit.

Indeed, research suggests that challenging media stereotypes of science isn't easy. When Steinke and her colleagues tested the impact of discussion-based media literacy interventions among a group of seventh graders, they found no effects on the students' attitudes toward science or their perceptions of women in science.[51] Nor did the interventions influence whether the students drew male or female scientists.[52]

Such findings suggest that education, by itself, can't fully counter the effects of media messages on young people's images of science. That's why experts such as Steinke and Long and organizations such as the Geena Davis Institute advocate making science-themed television programs more diverse.[53] The development of shows such as PBS's *SciGirls*, CBS's *Mission Unstoppable*, and Netflix's *Brainchild* offer grounds for optimism on this front, but the entertainment industry still has much room for improvement when it comes to presenting an inclusive portrait of science. We enjoyed watching *The Powerpuff Girls, Kim Possible,* and *Phineas and Ferb* with our own children, but we hope to see more scientists like Princess Bubblegum, and Doc McStuffins in the future.

Looking back over our case studies, it's also clear how demographic stereotypes of science cut across many different types of media. Depictions of strange or even villainous scientists may be on the wane, but the standard image of the white, male, heterosexual, cisgender, able-bodied scientist lives on in Hollywood movies, prime-time sitcoms, YouTube videos, and animated children's shows. Given the breadth of this pattern, promoting a more diverse vision of science will require a range of approaches and participants. With that in mind, let's consider how three groups—scientists, media producers, and audience members—can not only help foster inclusivity in science but also build support for scientific endeavors, bridge gaps between scientists and laypeople, and encourage critical engagement with science.

Notes

1 "Bill Nye: I took astronomy from Carl Sagan," *Secret Life of Scientists and Engineers*, PBS, March 21, 2014, www.pbs.org/video/secret-life-scientists-bill-nye-i-took-astronomy-carl-sagan.
2 Boss, Kitt, "The Bill Nye effect," *Seattle Times*, December 18, 1994, https://archive.seattletimes.com/archive/?date=19941218&slug=1948193.
3 Boss, "The Bill Nye effect."
4 See the Appendix for details.
5 Besley, John C., "Predictors of perceptions of scientists: Comparing 2001 and 2012," *Bulletin of Science, Technology & Society* 35, no. 1–2 (2015): 3–15; Losh, Susan Carol, "Stereotypes about scientists over time among US adults: 1983 and 2001," *Public Understanding of Science* 19, no. 3 (2010): 372–382; National Science Board, "The state of U.S. science and engineering 2020," National Science Foundation/National Science Board, 2020. https://ncses.nsf.gov/pubs/nsb20201.
6 See Chapter 1.
7 Long, Marilee, and Jocelyn Steinke, "The thrill of everyday science: Images of science and scientists on children's educational science programmes in the United States," *Public Understanding of Science* 5, no. 2 (1996): 101–120; Long, Marilee, Jocelyn Steinke, Brooks Applegate, Maria Knight Lapinski, Marne J. Johnson, and Sayani Ghosh, "Portrayals of male and female scientists in television programs popular among middle school-age children," *Science Communication* 32, no. 3 (2010): 356–382; Ryan, Lisa, and Jocelyn Steinke, "'I want to be like...': Middle school students' identification with scientists on television," *Science Scope* 34, no. 1 (2010): 44–49; Steinke, Jocelyn, Brooks Applegate, Maria Lapinski, Lisa Ryan, and Marilee Long, "Gender differences in adolescents' wishful identification with scientist characters on television," *Science Communication* 34, no. 2 (2012): 163–199; Steinke, Jocelyn, Maria Knight Lapinski, Nikki Crocker, Aletta Zietsman-Thomas, Yaschica Williams, Stephanie Higdon Evergreen, and Sarvani Kuchibhotla, "Assessing media influences on middle school-aged children's perceptions of women in science using the Draw-A-Scientist Test (DAST)," *Science Communication* 29, no. 1 (2007): 35–64; Steinke, Jocelyn, Maria Lapinski, Marilee Long, Catherine Van Der Maas, Lisa Ryan, and Brooks Applegate, "Seeing oneself as a scientist: Media influences and adolescent girls' science career possible selves," *Journal of Women and Minorities in Science and Engineering* 15, no. 4 (2009): 270–301.
8 Bandura, Albert, "Human agency in social cognitive theory," *American Psychologist* 44, no. 9 (1989): 1175–1184; Bandura, Albert, "Social cognitive theory of mass

communication," *Media Psychology* 3, no. 3 (2001): 265–299; Long and Steinke, "The thrill of everyday science."

9 Steinke, Jocelyn, "A portrait of a woman as a scientist: Breaking down barriers created by gender-role stereotypes," *Public Understanding of Science* 6 (1997): 409–428; Steinke, Jocelyn, "Women scientist role models in television programming," *Journal of Broadcasting & Electronic Media* 42, no. 1 (1998): 142–151.

10 Ryan and Steinke, "'I want to be like'"; Steinke et al., "Seeing oneself as a scientist."

11 LaFollette, Marcel Chotkowski, *Science on American television: A history*, University of Chicago Press, 2013, 174.

12 See the Appendix for details.

13 LaFrance, Adrienne, "The kids' show that taught me to ask, 'Why?'" *Atlantic*, April 30, 2016, www.theatlantic.com/entertainment/archive/2016/04/ode-to-3-2-1-contact/480546.

14 LaFrance, "The kids' show that taught me."

15 LaFollette, *Science on American television*; Mendoza, N. F., "'PBS' science project: *Newton's Apple* begins its 10th year of making learning also interesting," *Los Angeles Times*, October 25, 1992, www.latimes.com/archives/la-xpm-1992-10-25-tv-1206-story.html.

16 Boss, "The Bill Nye effect"; LaFollette, *Science on American television*.

17 Moore, Scott, "The madcap scientist of *Beakman's World*," *Washington Post*, September 19, 1993, www.washingtonpost.com/archive/lifestyle/tv/1993/09/19/the-madcap-scientist-of-beakmans-world/42a49f7b-32de-438a-a02d-b01d91b40582.

18 Gupta, Anita, "Meet Mr. Wizard, television's original science guy," *Smithsonian Magazine*, August 26, 2015, www.smithsonianmag.com/smithsonian-institution/meet-mr-wizard-science-guy-inspired-bill-nye-180956371.

19 See the Appendix for details.

20 Ryan and Steinke, "'I want to be like.'"

21 Long et al., "Portrayals of male and female scientists."

22 "DragonflyTV," Twin Cities PBS, January 28, 2021, www.tpt.org/dragonfly-tv/about-dragonflytv.

23 Affo, Marina, "Wilmington's STEM Queen is now a series regular on CBS show *Mission Unstoppable*," *Delaware News Journal*, January 5, 2021, www.delawareonline.com/story/news/2021/01/05/wilmington-stem-queen-now-series-regular-cbs-mission-unstoppable/4128294001.

24 Garcia-Navarro, Lulu, "New Netflix show *Brainchild* makes science fun for kids," NPR, January 27, 2019, www.npr.org/2019/01/27/688810237/new-netflix-show-brainchild-makes-science-fun-for-kids.

25 Bandura, Albert, Dorothea Ross, and Sheila A. Ross, "Imitation of film-mediated aggressive models," *Journal of Abnormal and Social Psychology* 66, no. 1 (1963): 3–11.

26 Steinke, Jocelyn, "Cultural representations of gender and science: Portrayals of female scientists and engineers in popular films," *Science Communication* 27, no. 1 (2005): 27–63.

27 Hoffner, Cynthia, "Children's wishful identification and parasocial interaction with favorite television characters," *Journal of Broadcasting & Electronic Media* 40, no. 3 (1996): 389–402; Steinke et al., "Seeing oneself as a scientist."

28 Bem, Sandra Lipsitz, "Gender schema theory and its implications for child development: Raising gender-aschematic children in a gender-schematic society," *Signs: Journal of Women in Culture and Society* 8, no. 4 (1983): 598–616.

29 Steinke, "Women scientist role models"; Long et al., "Portrayals of male and female scientists."

30 Ruvolo, Ann Patrice, and Hazel Rose Markus, "Possible selves and performance: The power of self-relevant imagery," *Social Cognition* 10, no. 1 (1992): 95–124; Steinke et al., "Seeing oneself as a scientist."

31 Burke, Kevin, *Saturday morning fever: Growing up with cartoon culture*, Macmillan, 1998; Perlmutter, David. *America toons in: A history of television animation*, McFarland, 2014.

32 Hains, Rebecca C., "Inventing the teenage girl: The construction of female identity in Nickelodeon's *My Life as a Teenage Robot*," *Popular Communication* 5, no. 3 (2007): 191–213.

33 Long et al., "Portrayals of male and female scientists."

34 The median age was 19.

35 Steinke, Jocelyn, and Marilee Long, "A lab of her own? Portrayals of female characters on children's educational science programs," *Science Communication* 18, no. 2 (1996): 91–115.

36 Long, Marilee, Greg Boiarsky, and Greg Thayer, "Gender and racial counter-stereotypes in science education television: A content analysis," *Public Understanding of Science* 10, no. 3 (2001): 255–269.

37 Long et al., "Portrayals of male and female scientists."

38 O'Connell, Margaret, "Mad science for girls (and boys), part two: The mildly mad scientific world of *Phineas and Ferb*," *Sequential Tart*, October 29, 2012, www.sequentialtart.com/article.php?id=2316.

39 Sharma, Paarth R., "View: One of the main reasons why Indian-Americans are subjected to racial abuse in US," *Economic Times*, May 7, 2017, https://economictimes.indiatimes.com/nri/nris-in-news/indian-americans-in-us-find-themselves-in-an-increasingly-strange-situation/articleshow/58554714.cms?from=mdr.

40 Geena Davis Institute on Gender in Media, "Portray her: Representations of women STEM characters in media," 2018, https://seejane.org/research-informs-empowers/portray-her.

41 See Chapter 3.

42 Steinke et al., "Assessing media influences."

43 Steinke et al., "Seeing oneself as a scientist."

44 Steinke et al., "Gender differences in adolescents' wishful identification."

45 Bond, Bradley J., "Fairy godmothers > robots: The influence of televised gender stereotypes and counter-stereotypes on girls' perceptions of STEM," *Bulletin of Science, Technology & Society* 36, no. 2 (2016): 91–97.

46 One notable exception to the latter pattern is Princess Bubblegum from *Adventure Time*, who once had a romantic relationship with Marceline the Vampire Queen.

47 Ryan and Steinke, "'I want to be like,'" 49.

48 McDuffie Jr., Thomas E., "Scientists—geeks & nerds?" *Science and Children* 38, no. 8 (2001): 16–19.

49 Fermilab, "Who's the scientist? Seventh graders describe scientists before and after a visit to Fermilab," March 2, 2000, https://ed.fnal.gov/projects/scientists/index.html.

50 Quigley, Robert, "Trip to FermiLab teaches children not all scientists are beaker-toting male weirdos," *The Mary Sue*, June 23, 2010, www.themarysue.com/fermilab-scientist-pictures-children.

51 Steinke, Jocelyn, Maria Lapinski, Aletta Zietsman-Thomas, Paul Nwulu, Nikki Crocker, Yaschica Williams, Stephanie Higdon, and Sarvani Kuchibhotla, "Middle school-aged children's attitudes toward women in science, engineering, and technology and the effects of media literacy training," *Journal of Women and Minorities in Science and Engineering* 12, no. 4 (2006): 295–323.

52 Steinke et al., "Assessing media influences."

53 Steinke and Long, "A lab of her own?"; Long et al., "Portrayals of male and female scientists"; Geena Davis Institute on Gender in Media, "Portray her."

11

RESHAPING POPULAR IMAGES AND PUBLIC PERCEPTIONS

Trevor Noah: This is amazing. The first picture ever of a black hole. And, like everything that's come out about it has been incredible, right? The size is incredible. It is bigger than our entire solar system ... And I know, some of you are like, "Trevor, what's the big deal? I already knew what a black hole looks like." No, you see, that's the thing. You didn't. None of us did. All you knew is what Hollywood made up ... and it turns out they were basically right.

—*The Daily Show with Trevor Noah* (April 10, 2019)

No one algorithm or person made this image, it required the amazing talent of a team of scientists from around the globe and years of hard work to develop the instrument, data processing, imaging methods, and analysis techniques that were necessary to pull off this seemingly impossible feat.

—Katie Bouman (Facebook post, April 10, 2019)

On April 10, 2019, the researchers working on the Event Horizon Telescope Collaboration publicly unveiled the first photograph of a black hole. Their accomplishment immediately became one of the biggest science news stories of the year. Leading newspapers printed the image on their front pages, science media organizations such as *National Geographic* delved into the techniques behind it, and all three major cable news networks ran segments about it. In covering the story, a wide range of outlets—some traditional, others less so—used "new discovery" frames that treated the project's breakthrough with wonder.[1] For example, *The Daily Show* ran a segment in which host Trevor Noah expressed his amazement at the photograph. In keeping with his role as a late-night comedian,

DOI: 10.4324/9781003190721-11

he then added a joke about the image's resemblance to the poster for the 1979 Disney movie *The Black Hole*.

Media messages also spotlighted Katie Bouman, the MIT computer scientist who led the development of the algorithm that captured the photograph. When an official MIT account tweeted a picture of her at a computer, with her hands partly covering her smile, the image went viral on social media.[2] Politicians such as Kamala Harris and Alexandra Ocasio-Cortez retweeted it, as did celebrities such as Elizabeth Banks (of *The Hunger Games*) and science communicators such as Emily Calandrelli (of *Emily's Wonder Lab*). Many observers celebrated Bouman as a role model of a woman working in a male-dominated STEM field and compared her to previous trailblazers who were less heralded in their own day. For example, journalist Flora Graham tweeted a photograph of Bouman alongside one of Margaret Hamilton, the lead computer programmer for NASA's Apollo moon mission. Other media messengers likened Bouman to the NASA mathematicians portrayed in the 2016 biographical film *Hidden Figures*: Katherine Johnson, Mary Jackson, and Dorothy Vaughn.[3]

Yet even as these observers were lauding Bouman's contributions, an army of Internet trolls was targeting her with a sexist harassment campaign. They created fake Instagram and Twitter accounts under her name, posted disparaging memes about her on Reddit, and uploaded YouTube videos about her with titles such as "Woman Does 6% of the Work but Gets 100% of the Credit: Black Hole Photo."[4] Some of these cyberbullies alleged that she had stolen credit from Andrew Chael, another computer scientist who worked on the project. Chael, however, used his own Twitter account to push back against the trolls targeting Bouman. "[W]hile I appreciate the congratulations on a result that I worked hard on for years," he wrote on April 11, 2019, "if you are congratulating me because you have a sexist vendetta against Katie, please go away and reconsider your priorities in life." He then identified himself as a gay astronomer, highlighting another dimension of diversity in STEM.

The harassment of Bouman became a news story itself, with outlets such as NBC News, the *Washington Post*, the *Atlantic*, and the Verge running stories about the backlash against her.[5] Vox's Brian Resnick went on to suggest that the media should present "more images of women thriving in science" to counter the sexism and discrimination that often make women feel unwelcome in science.[6] At the same time, a few observers pointed out that the media's narrow focus on Bouman reinforced the cultural myth of the lone genius in science. Bouman herself wrote a Facebook post about how the Event Horizon Telescope Collaboration was a collective effort, while the *New York Times* quoted another scientist from the project, Sara Issaoun, who "warned against a 'lone-wolf success' narrative."[7] Similarly, Resnick described the history of the lone genius trope—which is more frequently applied to men, rather than women scientists—and explained how "in present times, science is hardly ever a solitary endeavor."[8]

The case of the black hole photograph illustrates two of this book's key themes. One is that contemporary media portrayals of science and scientists tend to be favorable, even glowing. The celebratory depictions of Bouman and her colleagues in media coverage echo the portrayals of heroic fictional scientists in hit movies and television dramas. Likewise, they parallel the positive depictions of real-world scientists in popular documentary programs, late-night comedy shows, and children's educational programs. In recent decades, "mad" or sinister scientists have been the exception, rather than the rule, in media portrayals of the profession.

Yet a second theme of our book is that media messages sometimes reinforce perceptions that can discourage audience members from engaging with science. Science fiction shows often present scientific work as dangerous, while depictions of scientists as strange and socially awkward live on in prime-time sitcoms and children's cartoons. Media ranging from blockbuster movies to YouTube videos also tend to stereotype scientists as white, heterosexual, able-bodied men. Furthermore, Internet trolls and cyberbullies frequently target science communicators with discriminatory attacks similar to the ones leveled at Bouman. Such messages create a challenge for efforts to foster a more diverse scientific community.

The evidence in this book highlights other challenges, too. Depending on their content, media messages can either help or hinder attempts to bridge divides between scientists and the public. For example, some news outlets and social media sites have reinforced the scientific consensus on issues such as climate change and COVID vaccinations whereas others have spread misinformation about the same topics. Nor is it an easy task for scientists and science communicators to reach the "missing audience" of people who are uninterested in science—or to persuade those who resist scientific information on the basis of their own worldviews.

Another ongoing challenge revolves around promoting critical engagement with and participation in science. Many efforts at science communication follow the top-down approach of the *deficit model*, which treats laypeople as passive vessels for information to be delivered through the media. This model suggests that the solution to any issue involving public awareness or action is simply for scientists to communicate more clearly and more often. By contrast, the *public understanding of science model* emphasizes the importance of considering how audience members actively draw on their own values and experiences in responding to media messages. In addition, this model highlights the benefits of promoting broad and critical-minded participation in science.

Now that we've examined how media messages portray science and how such portrayals influence public perceptions of science, it's time to consider the lessons we can draw from our findings. Specifically, let's look at what scientists, media producers, and audience members can do to help address the key challenges we've outlined. While we're at it, we'll discuss what researchers like us are still learning about science in the media and how we can develop new insights on the topic.

Lessons for Media Producers

Although many different types of media present messages about science, two broad communication principles apply across all of them. The first is that media producers—be they directors, screenwriters, journalists, comedians, or social media content creators—should begin by identifying and reflecting on their goals when they portray science and scientists. Some media producers strive to engage audience members, counter misinformation, encourage activism, or inspire a new generation of scientists. Others aim to draw box office returns, ratings, donors, followers, or subscribers. Still others hope to do both, and these two sets of goals can clash with or complement one another. Audience considerations sometimes lead media producers to offer depictions that fuel misperceptions and stereotypes of science, but a number of the cases we've looked at illustrate how media producers can appeal to audience members while simultaneously promoting science engagement and understanding.

A second general principle for those working in the media is to craft messages strategically and intentionally based on their communication goals. Here, it's important for media producers to understand how their fellow creators are portraying science as well as how their own messages may influence audience members. For example, writers and directors for Hollywood movies and prime-time television programs could weigh the role their depictions play in *cultivating* perceptions of science and scientists. Similarly, makers of documentary and reality television shows could take into account how their messages *prime* thoughts or fears in viewers' minds, just as journalists and social media content producers could consider the consequences of their choices about how to *frame* scientific topics. All of these producers, and especially makers of media aimed at children, could also draw on theories about media *models* to understand how their portrayals shape young viewers' aspirations toward scientific careers.

Having laid out these general guidelines, let's take a closer look at the specific lessons our findings suggest—starting with entertainment and infotainment media images of what scientists are like. If such depictions sometimes reinforce perceptions that the scientific community is full of oddballs like *The Big Bang Theory*'s Sheldon Cooper and *Phineas and Ferb*'s Dr. Doofenshmirtz, they can also cultivate images of scientists as likable, trustworthy people who work for the good of society. Recent movies and television shows have offered a variety of models in this mold, from fictional ones such as the NASA scientists of *The Martian* to real-world ones such as the young scientists on *Mission Unstoppable*. Entertainment and infotainment portrayals can also cultivate perceptions of science as a collaborative and welcoming endeavor that's open to everyone. For example, *MythBusters* presents a team of relatable special effects experts working together on fun, interesting scientific projects.

In shifting away from the old image of the "mad, bad scientist" epitomized by Dr. Frankenstein, media producers should take care not to fall into the opposite extreme: an excess of hero worship.[9] Media portrayals of scientists as "wizards" or "high priests" may reinforce uncritical deference to scientific authority along with perceptions of the scientific community as distant from everyone else, just as depictions of "lone geniuses" who solve problems through flashes of inspiration perpetuate cultural myths about how science works.[10] Meanwhile, images that humanize scientists may help viewers connect science to their everyday lives and see it as something in which they, too, can take part. The flawed but likable forensic scientists of *CSI* and the down-to-earth research teams featured in recent *NOVA* episodes offer examples of such portrayals in television dramas and documentaries, respectively.

Besides shaping general perceptions of science and scientists, entertainment and infotainment media can influence beliefs about topics ranging from astronomy to zoology. On the one hand, Hollywood depictions that create a sense of perceptual realism through visuals and language sometimes promote distorted understandings of specific scientific topics, as in the case of *The Core*'s bogus geology. On the other hand, flawed but convincing depictions of science can also serve constructive ends. Take *The Day After Tomorrow*: this disaster movie didn't get everything right about climatology, but its creators achieved their goals of making a box office hit *and* raising concern about climate change.

In light of such effects, entertainment media producers should consider the nuances of what *looks* and *sounds* believable compared to what *is* realistic. Some Hollywood scenarios, such as the mission to Mars in *The Martian*, not only seem but are plausible, whereas other scenarios, such as the revival of extinct dinosaurs in *Jurassic World*, make the implausible appear doable or even likely. Similarly, the forensic techniques that *CSI* and other crime dramas depict as swift and virtually infallible vary widely in their real-world reliability: trace analysis of substances such as hair is scientifically questionable, and even the "gold standard" technique in forensic science—DNA testing—is often slower in reality than on prime time. Of course, movie and television producers are typically more interested in evoking willing suspension of disbelief among audience members than in crafting fully accurate depictions of science. Yet consulting with scientists can help them balance entertainment and engagement within the context of their storytelling goals.

For their part, infotainment media programs can sow cynicism, promote misperceptions, and prime fears through sensationalized messages—or, alternatively, encourage science engagement through their dramatic (if oversimplified) storylines and vivid imagery. Critics have argued, with justification, that "docufictions" such as *Mermaids: The Body Found* erode trust in scientific experts, just as documentary programs such as *Shark Week* fan exaggerated worries about shark attacks and reality shows such as *Ghost Hunters* bestow spurious trappings of scientific legitimacy on paranormal investigators. Still, the makers and distributors of such programs could help mitigate at least some of these effects by working with experts on public service announcements (as *Shark Week* did for a while) and by featuring disclaimers that identify fictitious or unsubstantiated content (as some paranormal-themed programs have done). Meanwhile, programs that blend a stronger dose of information with their entertainment values can reach broad audiences while teaching viewers about scientific methods and reasoning. With its mix of explosions and experimentation, *MythBusters* provides one model of this approach.

News coverage of science, in turn, wields a double-edged impact when it comes to shaping perceptions of science-related issues. Such coverage can promote public acceptance of mainstream scientific conclusions, but it can also help exacerbate gaps between what scientists say and what the public believes. Given this, journalists should weigh the potential effects of how they frame issues involving science. For example, "runaway science" or "Frankenstein's monster" frames carry dramatic appeal but can fuel distorted views of emerging technologies such as nanotechnology and artificial intelligence.[11] Likewise, strategy frames emphasizing debate tactics and "postgame analysis" seem objective but can create false equivalence on issues such as evolution and global warming.

Building on this last point, journalists should reflect on how—and to whom—they bestow the mantle of scientific authority. "Balanced" coverage may help legitimize claims with no scientific support, such as those advanced by intelligent design proponents, climate change skeptics, and anti-vaxxers. By the same logic, news stories that present claims about UFOs or psychic powers without including expert rebuttals can lend a false aura of scientific credibility to paranormal researchers. When the weight of scientific evidence falls heavily on one side of an issue, reporters are *not* obligated to present both sides equally in the name of objectivity. Far from it: it's their responsibility to highlight the scientific consensus and debunk spurious claims. Some news outlets have taken proactive approaches to doing so on subjects such as climate change and COVID conspiracy theories, but others—including a number of Fox News talk shows—have fallen short on this score.[12]

Beyond traditional journalism, outlets that blend news and entertainment values through humor can continue—and even expand—their roles as gateways to science. Over the past two decades, late-night television hosts such as Stephen Colbert, John Oliver, and Samantha Bee have used satire to promote science engagement along with acceptance of the scientific consensus on issues such as climate change and vaccinations. In particular, these comedians have demonstrated how humorous messages about science can reach Americans who aren't particularly interested in or informed about the topic itself. Our findings also suggest that the power of humor isn't limited to a handful of late-night hosts. For example, tongue-in-cheek messages in news stories about ESP can help counter tenuous claims to scientific authority by paranormal researchers.

To be sure, media producers should use the tools of satire with care. Not every audience member will necessarily get every joke, particularly if it's an ironic one. Satirical comedy can also provide a platform for questionable scientific claims or dismissive mockery of scientific discoveries—as when Neil Young made spurious statements about GMOs on *The Late Show*, or when comedians on *The Nightly Show* mocked Bill Nye's enthusiasm about finding water on Mars.[13] Still, humor provides a promising path for debunking anti-science voices and challenging false balance in news coverage (as with *Last Week Tonight*'s "statistically representative climate debate") as well as for promoting critical thought about scientific methods and findings (as with the same program's segments on scientific studies and forensic science).

As for social media, they provide platforms for both traditional media producers (such as *National Geographic* and The Discovery Channel) and a host of new content creators to communicate about science. Unfortunately, sites such as Facebook, Twitter, Instagram, Reddit, and YouTube sometimes facilitate sensationalized clickbait, harmful misinformation, polarized discussions, and—as in the case of the harassment targeting Katie Bouman—sexist bullying. In a more positive development, social media companies have begun trying to counter scientific misinformation through fact-checking and algorithm-driven links to trustworthy sources. Furthermore, individual content creators can use the affordances provided by social media to foster science engagement through lighthearted "new discovery" frames, as IFLS's Elise Andrew has done, and to call out sexism in science, as *The Brain Scoop*'s Emily Graslie has done.

The latter point brings us back to a broader issue with media portrayals: they often reinforce demographic stereotypes of science, thereby sending the message that only white, heterosexual, able-bodied men can become scientists. On the plus side, we've seen how the media can present more diverse models for viewers to emulate. For example, two recent movies featuring Black women as scientists—*Black Panther* and *Hidden Figures*—found box office success and critical acclaim. In the world of television, *The X-Files* and *Bones* illustrate how the media industry can showcase women scientists who inspire wishful identification among viewers and encourage young women in particular to imagine "possible selves" as future scientists.[14] Meanwhile, programs such as *Cosmos* and *Brainchild* provide examples of how makers of documentary and educational television shows can help foster a more inclusive vision of science. Media producers could apply the same principles in depicting more LGBTQ scientists and disabled scientists, two groups rendered largely invisible by popular movies and television shows. Indeed, the media industry may need to *over*represent scientists from historically underrepresented groups if it hopes to counteract its long history of stereotypical messages.[15]

Yet it's also important to remember that media messages can sometimes produce complex effects. Take public perceptions of the gender gap in science: higher levels of television viewing go hand in hand with rosier views of this gap, perhaps because the gender imbalance among scientists on prime time is still more equitable than the gender imbalance in real life. Beneath the overall numbers, the specifics of media depictions matter, too. If women scientists in movies and television shows are overwhelmingly biologists rather than physicists or engineers, or if Asian scientists in the media tend to be nerdy, or if LGBTQ scientists and disabled scientists always play sidekick roles, then these portrayals may reinforce audience members' stereotypes. Such complexities require thoughtful navigation, but the benefits to society justify the extra work. Encouragingly, the evidence suggests that media producers can feature greater diversity in their portrayals of science without sacrificing popularity or plausibility among viewers.

Still, changes behind the scenes may be just as important as any changes in front of the camera. If the scientists *in* the media should reflect the diversity of our society, then so should the screenwriters, directors, journalists, satirists, and social media producers who *shape* the media. Such inclusivity on the creative side matters for what messages audiences receive. For example, the depiction of Shuri, the lead scientist in *Black Panther*, reflects not only the performance of the Black actor who plays her but also the vision of the film's Black director, Ryan Coogler, and his Black screenwriting partner, Joe Robert Cole. Likewise, the images of science in *Mission Unstoppable* reflect the work of the show's women-led creative team, including host/producer Miranda Cosgrove, producer Geena Davis, and showrunner Anna Wegner. These examples and others like them highlight how fostering diversity in the media industry can help foster diversity in the scientific community.

Lessons for Scientists

The same broad principles we've laid out for media producers apply to scientists and science communicators, as well. If they hope to engage audience members, bridge divides with the public, and recruit more diverse scientists, then they should choose media platforms and messages that match those aims. As we've seen, Americans learn about science from a wide range of media outlets, from Hollywood movies and prime-time sitcoms to late-night shows and social media. To reach their target audiences, particularly the "missing" members who tune out or reject messages about science, scientists and their allies should communicate through as many of these outlets as possible—so long as the outlet fits their goals. The media effects theories we've explored offer guidance for how to do so.

Not every scientist needs to go to Hollywood, of course, but the ones who land opportunities as consultants can influence the entertainment industry's portrayals of science and scientists. For example, the NASA scientists who worked with the makers of *The Martian* helped produce a plausible and engaging depiction of a mission to another planet—a real-world goal of the space agency. Likewise, consultants for television shows can help shape how the medium cultivates perceptions of science and help develop media models who inspire future scientists: consider the "Scully effect" on young women who watched *The X-Files* and the potential "*CSI* effect" on student interest in forensic science. Media producers and audience members may tend to see plausibility and entertainment value as more important than accuracy, but scientists who take on roles as consultants can work within these constraints to make a difference.

Scientists who engage with documentary and educational programs can draw parallel lessons. In some cases—as with *Cosmos*, *Nova*, *MythBusters*, and *Bill Nye the Science Guy*—they'll find chances to help foster perceptions of science as an exciting endeavor where people from many different backgrounds and walks of life work together to benefit society. Such shows can also offer platforms for promoting a greater understanding of scientific methods and reasoning along with a less monolithic and more participatory vision of science itself. In other cases, scientists may want to use transmedia strategies to challenge or contextualize infotainment depictions that promote misperceptions, prime fears, and undermine trust in scientific institutions. Andrew David Thaler and David Shiffman's effort to debunk two Discovery Channel docufictions, *Megalodon* and *Shark of Darkness*, provides one roadmap for using social media messaging and search engine optimization to counter distorted portrayals of science.

When communicating with the news media, scientists should be aware of how they're framing the issue at hand—as well as how journalists frame science. Many reporters seek out angles that resonate with news values such as drama and novelty, and all reporters need to tell stories that audience members will understand. If scientists deride framing as a public relations gimmick and call for news outlets to "just stick to the facts," then they'll be taking a self-defeating approach given that frames are a fundamental part of communication by reporters *and* scientists themselves. A more constructive approach would be to follow recommendations from communication researchers and science communicators about effective framing strategies. For instance, Jason Rosenhouse and Glenn Branch advise scientists discussing evolution in the media to be ready to rebut frames that call for "fairness" or for teaching "both sides."[16] On the issue of climate change, Matthew Nisbet recommends that science communicators reframe their messages to broaden support for action and overcome value-driven resistance; in particular, he suggests emphasizing economic and public health benefits from addressing climate change.[17] Likewise, Zeynep Tufekci argues that public health experts should learn from past mistakes in framing public health issues such as

COVID-19: she warns against overemphasizing the limitations of vaccines and recommends doing more to highlight their social benefits.[18]

In addition to talking with traditional journalists, scientists can reach audience members—including less engaged ones—through humorous outlets. Comedy-focused media such as *The Late Show* and *Full Frontal* frequently offer scientists opportunities to speak at length, rather than in brief soundbites, about their work. Moreover, humorous outlets and satirical messages can help debunk misinformation and reinforce the scientific consensus on topics ranging from climate change and vaccines to ESP—which, in turn, can shift public perceptions. This doesn't mean that scientists themselves always need to be funny; sometimes playing along with experienced comedians is enough to get the message out to audience members.

Compared to other sorts of media, social media platforms stand out for how they give scientists affordances for communicating directly with the public through their own words and images. In using these platforms, scientists should keep in mind that different platforms offer different affordances. Scientists may want to use Twitter to share information about their research and their own lives, as astronomers such as Neil deGrasse Tyson and Pamela Gay have done, while they can use Reddit to engage in conversations with laypeople through science-themed subreddits and "Ask Me Anything" threads. They can also use social media posts or comments to help correct misinformation on important issues and hashtags such as #distractinglysexy to challenge discrimination in their own profession. Furthermore, they can use image-sharing social media platforms to humanize scientists and counter stereotypes of who can become one. The "Scientists Who Selfie" project shows that women scientists who post pictures of themselves on Instagram can help promote perceptions of scientists as relatable and trustworthy while simultaneously dispelling perceptions of science as a "male" activity.[19]

As social media sites evolve and new sites emerge, scientists should adapt their communication strategies accordingly. Consider TikTok: the surge in popularity of this short-form video platform during the late 2010s and early 2020s created new opportunities for scientists to communicate with its young-skewing user base. For example, Darrion Nguyen created his account, @lab_shenanigans, for fun but used its growing popularity to promote engagement with chemistry and "'show that not all scientists are male and white.'"[20] As of February 2021, he had more than half a million followers. Another "science star" of TikTok is bioengineer Anna Blakney (@anna.blakney), who has posted dance videos along with videos debunking misinformation about COVID-19 vaccines. One of her videos, in which she pretends to spill a vial of vaccine, currently has more than 16 million views.[21] These accounts and many others illustrate how scientists can use new platforms such as TikTok to humanize themselves and make their work more accessible to laypeople.

None of this is to say that communication through the media the only way for scientists to engage with the public. If "parasocial" contact with scientists in the media can provide role models for young people to follow and help break down stereotypes of the profession, then so can interpersonal contact in the form of class visits or field trips. Likewise, personal engagement by scientists and other experts can influence perceptions of issues such as climate change both directly and indirectly through a "two-step flow of information" where messages circulate among members of the public through conversations.[22] In short, scientists should view media outreach as one tool among many for communicating with broader audiences.

Lessons for Members of the Public

Just as media producers and scientists can draw lessons from our findings, so, too, can members of the public—including our readers. By understanding how media messages portray science and cultivate perceptions, each of us can become a more informed and critical consumer of such messages. For example, familiarity with media tropes about scientific work and media stereotypes of scientists can help us to recognize such elements in the movies and television programs we watch. When science fiction films and shows portray research projects that unleash killer dinosaurs or summon deadly aliens, we can keep in mind that real-world science is seldom so dangerous. And when sitcoms and children's cartoons present scientists who are socially awkward geeks, we can see them in the context of a history that includes the original *Nutty Professor, Back to the Future*'s Doc Brown, and *The Big Bang Theory*'s Sheldon Cooper.

On a more "nuts and bolts" level, learning about the visual and sound techniques of Hollywood movies, prime-time television, and infotainment documentaries can help us recognize these techniques in action. For example, we can spot how the superhero movies of the Marvel Cinematic Universe create an impression of plausibility through elaborate sets, computer-generated imagery (CGI), and impressive-sounding dialogue about "quantum entanglement," or how nature-themed programs such as *Shark Week* prime fears through ominous music, slow-motion camera work, and shots of blood in the water. We can also catch when forensic crime dramas such as *CSI* glamorize—and compress—laboratory work through cool blue lighting, propulsive music, and rapid editing, or when paranormal investigators like the ones on *Ghost Hunters* try to dress up their research as scientific by using EMF detectors and talking about "EVPs."

Similarly, understanding framing theory can prepare us to identify and weigh common frames in news coverage and social media messages. On any given scientific topic, media messengers may present many different storylines. Each of these frames emphasizes a particular interpretation of the issue—and, in doing so, deemphasizes other possible interpretations. Thus, we should ask ourselves not only what the frame at hand *suggests* but also what it *omits*. For example, does casting GMO foods as "Frankenfoods" hype fears while neglecting scientists' assurances that such foods are safe to eat? Likewise, do stories highlighting a small number of adverse reactions to COVID-19 vaccines heighten risk perceptions while downplaying the benefits of such vaccines? We should be particularly wary of frames, such as the "teach the controversy" frame for evolution and the "unsettled science" frame for climate change, that understate an existing scientific consensus on the issue in question.

Turning to media models, what we've learned about social cognitive theory may help us resist messages that discourage our aspirations and help us find models that reinforce more inclusive visions of who we can become. When we see the "standard image" of the scientist as a white man in a white lab coat, we can recognize it for what it is and then look beyond it to broader possibilities. Furthermore, we can seek out portrayals that help us imagine new possible selves and that nurture wishful identification with scientists. Some audience members may look to real-life media models such as Katie Bouman and *Mission Unstoppable*'s Jacqueline "the STEM Queen" Means. Others may draw inspiration from fictional characters such as Shuri from *Black Panther*, Temperance Brennan from *Bones*, or Cisco Ramon from *The Flash*. Still others may see their potential future in cartoon scientists such as Ms. Frizzle from *The Magic School Bus* or Princess Bubblegum from *Adventure Time*.

Yet reflecting on media messages and media effects is only one step toward positive change. Audience members who want to take a more active role in science communication can draw further guidance from our evidence on how to challenge harmful messages and reinforce constructive ones. For example, we can correct misinformation about topics such as climate change and COVID-19 vaccines when we see it on Facebook, Twitter, or Reddit. Likewise, we can use social media posts and hashtags to call out sexist or racist internet bullying of science communicators. We also can share lighthearted "new discovery" frames that promote science engagement, as well as satirical humor that encourages critical thought about scientific messages. In the offline world, we can raise alternative interpretations when family members and friends cite dramatized or misleading media frames for scientific topics. And those of us who are caregivers for children can introduce more diverse media models of scientists.

Having said all this, individualistic approaches to countering distorted or stereotypical media messages are ultimately limited in what they can do. As audience members, we simply can't resist every media message we encounter. We face too many competing demands on our attention to maintain a constant guard against media influence, and we receive too many messages to scrutinize— let alone actively challenge—them all.[23] Compounding our dilemma, many media messages include features that enhance their power to bypass critical examination. Perceptually realistic portrayals of fictitious science can induce our willing suspension of disbelief, frames that resonate with our worldviews can carry extra weight in our judgments, humor can disarm our defenses, and media stereotypes can fit readily into mental structures we developed during childhood.

Nor is critical thought a cure-all for media influence. In some cases, it even plays the opposite role: media messengers can and do exploit our tendency toward motivated reasoning as a tool for promoting distorted understandings of science. Consider how politicians and interest groups often publicize frames that help audience members rationalize positions founded on ideological or religious beliefs. For example, creationists and "intelligent design" proponents have argued for "teaching the controversy" on evolution, and climate change skeptics have claimed that "the science isn't settled" enough to justify policy action. Similarly, proponents of conspiracy theories on topics from cryptozoology to COVID-19 vaccines have taken advantage of motivated reasoning by appealing to—and potentially reinforcing—cynicism toward scientific agencies such as NOAA and the CDC. Because we often see and hear what we want to in media messages, exposure to more scientific information doesn't necessarily translate into greater acceptance of what the science says on any particular topic. Our beliefs can even shape how we interpret satire about science, as with Stephen Colbert's ironic climate-themed humor.

Efforts such as public service announcements and media literacy training programs can help educate audience members, but they're not surefire solutions in every situation. Witness *Shark Week*'s public service announcements from scientific experts, which may have boosted support for shark conservation efforts but didn't dispel exaggerated fears about sharks. Similarly, classroom interventions may help students recognize media stereotypes of scientists but won't always neutralize their effects on perceptions.[24]

In the end, any fundamental shifts in the links between media portrayals of science and public perceptions will require systemic changes in what messages and messengers we see, along with who creates the messages and chooses the messengers. So, what role can laypeople play in fostering such changes? To start with, we can support media that present engaging and diverse portrayals of scientists, scientific reasoning, and scientific methods—especially when they're produced by diverse creators. In some cases, this could take the form of watching movies or television shows to boost their viability. On social media platforms, it could include liking posts, sharing videos, and subscribing to channels. We can also use social media to organize online *and* offline collective action that challenges media misinformation or stereotypes and that advocates for different content in the future.[25]

Going beyond our role as audience members for messages about science, we can become more active in science itself. The public understanding of science model highlights the potentially transformative effects of citizen participation in each step of the scientific process, from designing research projects to making decisions based on their results. Such "citizen science" can take many forms, including crowdsourced studies where citizens participate through websites or mobile apps.[26] In addition, this model highlights the value of broader citizen involvement in science communication—particularly in terms of fostering two-way dialogue between scientists and laypeople rather than top-down transmission of information. Despite their built-in limitations and the toxic behavior of some users, social media platforms may be especially powerful tools for promoting citizen science: they provide affordances for communication about findings as well as consultation with and participation by laypeople.[27]

Extending the public understanding of science model to our own field, we believe it's important to encourage citizen participation in research *about* science communication, as well. In fact, we foresee many opportunities for citizen science on the topics we've raised in this book.

Ongoing Challenges for Science Communication Researchers

Though the theories and evidence we've described shed light on many aspects of how media messages present science and how such messages influence audience members, we still have puzzles left to solve. For example, we know a fair amount about how movies and television shows depict scientists in terms of gender and race, but less about the impact of such depictions—and even less when it comes to portrayals of LGBTQ scientists, disabled scientists, or scientists who face "double binds" from intersecting barriers to the profession. We also know less about how diversity—or a lack thereof—in social media messages about science influences audience members' perceptions of the profession.

In addition, science communication research could benefit from greater attention toward audience members as active participants rather than passive recipients of messages. Most of the evidence we've presented in this book comes from content analysis, surveys, and experiments. However, these methods provide only one set of windows into public perceptions and behavior regarding science. We know from in-depth interviews, focus group research, and participant observation that laypeople think, talk, and act in complex ways when it comes to the subject: they draw on their own experiences and worldviews along with media messages to make sense of science and their own relationships with it.[28] More research that combines quantitative and qualitative methods would offer us a richer portrait of how audiences respond to media messages about science. Such research could be especially useful in helping us better understand how to foster broader citizen engagement with science and science communication.

A final point to consider is that our evidence provides snapshots of media messages about science and their effects during one time period. Science is constantly changing, and so is the media landscape; as a result, research on the links between the two must evolve as well. Over the past decade, we've witnessed a transformation of the public's movie and television viewing habits through the rise of streaming video services such as Netflix and Disney Plus. Likewise, we've seen the collapse of some popular social media platforms and the explosive growth of others, such as Twitter, Instagram, and, more recently, TikTok. We've also experienced broader social shifts and events that have shaped both audience values and the media environment, including deepening political polarization, the emergence of social movements such as #MeToo and Black Lives Matter that highlight ongoing harassment and systemic discrimination, and a pandemic that placed medical science at the center of public attention while altering many aspects of daily life. The theories we've explored in this book can help us understand media portrayals of science and their effects through such developments, but some of their applications and their implications will inevitably change with the times.

For our part, we look forward to the future of science in the media with cautious optimism. Many of the old patterns we've discussed will undoubtedly continue through the 2020s and beyond. Movies will depict unlikely scientific scenarios, entertainment television shows will mine the dramatic potential of dangerous science, and infotainment programming will hype fears and cynicism through sensationalism. News coverage will cast emerging technologies as Frankenstein's monsters or Pandora's boxes and give airtime to questionable claims in the name of balancing "both sides," while misinformation and identity-based cyberbullying will persist on social media. Yet we're hopeful that new generations of media producers will reimagine science as a more inclusive community; that new generations of scientists will find creative ways to engage laypeople through media platforms; and that new generations of audience members will challenge future science communicators to do better. We're also hopeful that our students and our readers can help make these changes happen.

Notes

1 Hitlin, Paul, and Kenneth Olmstead, "The science people see on social media," Pew Research Center, March 21, 2018, www.pewresearch.org/science/2018/03/21/the-science-people-see-on-social-media.

2 Mervosh, Sarah, "How Katie Bouman accidentally became the face of the Black Hole Project," *New York Times*, April 11, 2019, www.nytimes.com/2019/04/11/science/katie-bouman-black-hole.html.

3 Resnick, Brian, "Male scientists are often cast as lone geniuses. Here's what happened when a woman was," *Vox*, April 16, 2019, www.vox.com/science-and-health/2019/4/16/18311194/black-hole-katie-bouman-trolls; Willingham, A.J., "We cheer on women in the sciences, but recruiting and retaining them is still a different story," CNN, April 12, 2019, www.cnn.com/2019/04/12/us/scientist-women-retention-sci-trnd/index.html.

4 Collins, Ben, "The first picture of a black hole made Katie Bouman an overnight celebrity. Then internet trolls descended," NBC News, April 12, 2019, www.nbcnews.com/tech/tech-news/first-picture-black-hole-made-katie-bouman-overnight-celebrity-then-n994081; Elfrink, Tim, "Trolls hijacked a scientist's image to attack Katie Bouman. They picked the wrong astrophysicist," *Washington Post*, April 12, 2019, www.washingtonpost.com/nation/2019/04/12/trolls-hijacked-scientists-image-attack-katie-bouman-they-picked-wrong-astrophysicist; Griggs, Mary Beth, "Online trolls are harassing a scientist who helped take the first picture of a black hole," *The Verge*, April 13, 2019, www.theverge.com/2019/4/13/18308652/katie-bouman-black-hole-science-internet; Koren, Marina, "The dark saga of Katie Bouman: How a young scientist got sucked into the black hole of the internet," *Atlantic*, April 15, 2019, www.theatlantic.com/science/archive/2019/04/katie-bouman-black-hole/587137; Resnick, "Male scientists."

5 Collins, "The first picture of a black hole"; Elfrink, "Trolls hijacked a scientists' image"; Griggs, "Online trolls"; Koren, "The dark saga."

6 Resnick, "Male scientists."

7 Mervosh, "How Katie Bouman became the face of the Black Hole Project."

8 Resnick, "Male scientists."

9 Haynes, Roslynn D., "Whatever happened to the 'mad, bad' scientist? Overturning the stereotype," *Public Understanding of Science* 25, no. 1 (2016): 31–44.

10 Gross, Rachel E., "*The Martian* and the cult of science," *Slate*, October 1, 2015, www.slate.com/articles/technology/future_tense/2015/10/ridley_scott_s_the_martian_film_science_worship_and_the_scientist_as_hero.html; Hornig, Susanna, "Television's *NOVA* and the construction of scientific truth," *Critical Studies in Media Communication* 7, no. 1 (1990): 11–23.

11 Bingaman, James, Paul R. Brewer, Ashley Paintsil, and David C. Wilson, "'Siri, show me scary images of AI': Effects of text-based frames and visuals on support for artificial intelligence," *Science Communication* 43, no. 4 (2021); Scheufele, Dietram A., and Bruce V. Lewenstein, "The public and nanotechnology: How citizens make sense of emerging technologies," *Journal of Nanoparticle Research* 7, no. 6 (2005): 659–667.

12 Feldman, Lauren, Edward W. Maibach, Connie Roser-Renouf, and Anthony Leiserowitz, "Climate on cable: The nature and impact of global warming coverage on Fox News, CNN, and MSNBC," *International Journal of Press/Politics* 17, no. 1 (2012): 3–31; Romer, Daniel, and Kathleen Hall Jamieson, "Conspiracy theories as barriers to controlling the spread of COVID-19 in the US," *Social Science & Medicine* 263 (2020): 113356.

13 Abad-Santos, Alex, "Watch: the segment that made me stop watching *The Nightly Show*," *Vox*, August 16, www.vox.com/2016/8/16/12502896/nightly-show-canceled-bad-roundtable.

14 21st Century Fox, Geena Davis Institute on Gender in Media, and J. Walter Thompson Intelligence, "The 'Scully effect': I want to believe … in STEM," Geena Davis Institute on Gender in Media, 2019, https://seejane.org/wp-content/uploads/x-files-scully-effect-report-geena-davis-institute.pdf; Gonzalez, Sandra, "How *Bones* bred a new generation of female scientists," CNN Entertainment, March 27, 2017, www.cnn.com/2017/03/27/entertainment/bones-tv-show-women-stem/index.html; Ryan, Lisa, and Jocelyn Steinke, "'I want to be like...': Middle school students' identification with scientists on television," *Science Scope* 34, no. 1 (2010): 44–49; Steinke, Jocelyn, Maria Lapinski, Marilee Long, Catherine Van Der Maas, Lisa Ryan, and Brooks Applegate, "Seeing oneself as a scientist: Media influences and adolescent girls' science career possible selves," *Journal of Women and Minorities in Science and Engineering* 15, no. 4 (2009): 270–301.

15 Long, Marilee, Jocelyn Steinke, Brooks Applegate, Maria Knight Lapinski, Marne J. Johnson, and Sayani Ghosh, "Portrayals of male and female scientists in television

programs popular among middle school-age children," *Science Communication* 32, no. 3 (2010): 356–382.

16 Rosenhouse, Jason, and Glenn Branch, "Media coverage of 'intelligent design,'" *BioScience* 56, no. 3 (2006): 247–252.

17 Nisbet, Matthew C., "Communicating climate change: Why frames matter for public engagement," *Environment: Science and Policy for Sustainable Development* 51, no. 2 (2009): 12–23.

18 On the Media, "The perils of pandemic doomsaying (and other COVID messaging mix-ups)," WNYC Studios, January 29, 2021, www.wnycstudios.org/podcasts/otm/segments/perils-pandemic-doomsaying-and-other-covid-messaging-mix-ups-on-the-media.

19 Jarreau, Paige Brown, Imogene A. Cancellare, Becky J. Carmichael, Lance Porter, Daniel Toker, and Samantha Z. Yammine, "Using selfies to challenge public stereotypes of scientists," *PloS one* 14, no. 5 (2019): e0216625.

20 Lemonick, Sam, "Chemists are finding their place on TikTok," *Chemical & Engineering News*, February 21, 2020, https://cen.acs.org/education/science-communication/Chemists-finding-place-TikTok/98/i8.

21 Trujillo, Anne, "Colorado native shares COVID-19 research using TikTok," *Denver Channel*, February 9, 2021, www.thedenverchannel.com/news/local-news/colorado-native-shares-covid-19-research-using-tiktok.

22 Nisbet, Matthew C., and John E. Kotcher, "A two-step flow of influence? Opinion-leader campaigns on climate change," *Science Communication* 30, no. 3 (2009): 328–354.

23 Eagly, Alice H., and Shelly Chaiken, *The psychology of attitudes*, Harcourt Brace Jovanovic, 1993; Petty, Richard E., and John T. Cacioppo, *Communication and persuasion: Central and peripheral routes to attitude change*, Springer, 2012.

24 Steinke, Jocelyn, Maria Knight Lapinski, Nikki Crocker, Aletta Zietsman-Thomas, Yaschica Williams, Stephanie Higdon Evergreen, and Sarvani Kuchibhotla, "Assessing media influences on middle school–aged children's perceptions of women in science using the Draw-A-Scientist Test (DAST)," *Science Communication* 29, no. 1 (2007): 35–64.

25 Ley, Barbara L., and Paul R. Brewer, "Social media, networked protest, and the March for science," *Social Media+ Society* 4, no. 3 (2018): 2056305118793407.

26 Bonney, Rick, Tina B. Phillips, Heidi L. Ballard, and Jody W. Enck, "Can citizen science enhance public understanding of science?" *Public Understanding of Science* 25, no. 1 (2016): 2–16; Lewenstein, Bruce V., "Can we understand citizen science?" *Journal of Science Communication* 15, no. 1 (2016).

27 Hargittai, Eszter, Tobias Füchslin, and Mike S. Schäfer, "How do young adults engage with science and research on social media? Some preliminary findings and an agenda for future research," *Social Media+ Society* 4, no. 3 (2018): 2056305118797720.

28 Bates, Benjamin R., John A. Lynch, Jennifer L. Bevan, and Celeste M. Condit, "Warranted concerns, warranted outlooks: A focus group study of public understandings of genetic research," *Social Science & Medicine* 60, no. 2 (2005): 331–344; Bauer, Martin, and Ingrid Schoon, "Mapping variety in public understanding of science," *Public Understanding of Science* 2 (1993): 141–155; Gamson, *Talking politics*.

APPENDIX: SURVEYS

The analyses original to this book used data from the following surveys of adult U.S. residents.

Center for Political Communication Survey (2016)

This survey was sponsored by the University of Delaware Center for Political Communication and conducted by Princeton Survey Research Associates International. A nationally representative sample of 900 respondents were interviewed by telephone on July 6–13, 2016.

Cooperative Congressional Election Survey (2016)

This survey was sponsored by the University of Delaware Center for Political Communication as part of a larger project conducted by YouGov. A nationally representative sample of 1,000 respondents were interviewed online on September 28–November 2, 2016. In addition, 855 of these respondents were re-interviewed online on November 9–December 13, 2016.

Cooperative Congressional Election Survey (2018)

This survey was sponsored by the University of Delaware Center for Political Communication as part of a larger project conducted by YouGov. A nationally representative sample of 1,000 respondents were interviewed online on September 27–November 5, 2018. In addition, 836 of these respondents were re-interviewed online on November 7–December 3, 2018.

Center for Political Communication Survey (2020)

This survey was sponsored by the University of Delaware Center for Political Communication and conducted by Qualtrics. The 1,052 respondents were sampled from a national Qualtrics panel and interviewed online on May 6–8, 2020.

INDEX

Printed in the United States
by Baker & Taylor Publisher Services